Relative Contractor Risks

A Data-Analytic Approach to Early Identification

PHILIP S. ANTON, WILLIAM SHELTON, JAMES RYSEFF, STEPHEN B. JOPLIN,
MEGAN MCKERNAN, CHAD J. R. OHLANDT, SAMANTHA COHEN

Prepared for the Department of the Air Force
Approved for public release; distribution unlimited

PROJECT AIR FORCE

For more information on this publication, visit **www.rand.org/t/RRA433-1**.

About RAND

The RAND Corporation is a research organization that develops solutions to public policy challenges to help make communities throughout the world safer and more secure, healthier and more prosperous. RAND is nonprofit, nonpartisan, and committed to the public interest. To learn more about RAND, visit www.rand.org.

Research Integrity

Our mission to help improve policy and decisionmaking through research and analysis is enabled through our core values of quality and objectivity and our unwavering commitment to the highest level of integrity and ethical behavior. To help ensure our research and analysis are rigorous, objective, and nonpartisan, we subject our research publications to a robust and exacting quality-assurance process; avoid both the appearance and reality of financial and other conflicts of interest through staff training, project screening, and a policy of mandatory disclosure; and pursue transparency in our research engagements through our commitment to the open publication of our research findings and recommendations, disclosure of the source of funding of published research, and policies to ensure intellectual independence. For more information, visit www.rand.org/about/principles.

RAND's publications do not necessarily reflect the opinions of its research clients and sponsors.

Published by the RAND Corporation, Santa Monica, Calif.
© 2022 RAND Corporation
RAND® is a registered trademark.

Library of Congress Cataloging-in-Publication Data is available for this publication.

ISBN: 978-1-9774-0875-4

Cover: Senior Airman Taylor Crul/U.S. Air Force; Astibuag/Adobe Stock.

Preface

As all acquisition professionals are aware, there are multiple kinds of risks in developing and acquiring new systems. One recurring challenge to successful acquisition program execution is poor contractor performance. When contractors are in danger of not meeting contractual performance goals, Department of the Air Force (DAF) acquisitions may not be fully aware of the shortfall until, for example, a schedule deadline is missed, government testing indicates poor performance, or costs exceed expectations.

This report presents a new way to apply data analysis on a variety of government and external data sources to assess the relative contractor performance risk in Air Force acquisition contracts and programs. This method seeks to produce risk indicators earlier than what current information sources and metrics can provide. This is the final report for Phase I of an effort to test this outlined approach by building a prototype that uses actual data to calculate relative risk measures and present these results to users to assess their relevance and refine their management of potential risks. This report summarizes our findings, including a taxonomy of contractor risks, relevant data sources, risk measures and equations, and a prototype that implements the relative risk measures on actual data sources. Note that we are focusing in this work on the types of relative risks related to enabling factors for contractor performance as opposed to those related to the design and technology involved in the delivered product or service. These enabling factors are, perhaps, those that receive the least attention and are hard to characterize because they rely on various data sources that are hard to access and for which integration and analysis are required. Hence, that is our focus for applying data access and analysis in this project.

It is also worth noting that we are focusing on a contractor's performance relative to its peers. This approach would highlight to acquisition professionals whether a contractor of interest is performing significantly worse than others in assessed dimensions. The acquisition professional would then be able to focus appropriate management attention on this area based on its relevance to the program and the level of risk deemed acceptable to the government.

This research should be of interest to acquisition professionals and leadership interested in ways to improve acquisition performance through early identification of potential relative contractor risks to inform active program management and mitigation of risks. The prototype should be of interest to acquisition officials (from program managers to milestone decision authorities) to help them access more data in an easy-to-understand way so they can focus their limited time on areas that require increased management attention. This approach should be useful during any phase of the acquisition process—from the pre–Materiel Development Decision phase through disposal.

This research was commissioned by the Deputy Assistant Secretary for Acquisition Integration (SAF/AQX), Office of the Assistant Secretary of the Air Force for Acquisition,

Technology, and Logistics. It was conducted within the Resource Management Program of RAND Project AIR FORCE as part of a project on the Early Indicators of Relative Contractor Performance Risk for Air Force Acquisition (PA18H-R8A4). Research was conducted from February 2018 through August 2019.

RAND Project AIR FORCE

RAND Project AIR FORCE (PAF), a division of the RAND Corporation, is the Department of the Air Force's federally funded research and development center for studies and analyses, supporting both the United States Air Force and the United States Space Force. PAF provides the DAF with independent analyses of policy alternatives affecting the development, employment, combat readiness, and support of current and future air, space, and cyber forces. Research is conducted in four programs: Strategy and Doctrine; Force Modernization and Employment; Workforce, Development, and Health; and Resource Management. The research reported here was prepared under contract FA7014-16-D-1000.

Additional information about PAF is available on our website:

www.rand.org/paf/

This report documents work originally shared with the DAF on August 9, 2019. The draft report, issued on October 14, 2020, was reviewed by formal peer reviewers and DAF subject-matter experts.

Executive Summary

A recurring challenge to successful acquisition program execution is poor contractor performance. The Deputy Assistant Secretary for Acquisition Integration asked RAND Project AIR FORCE to develop and prototype data-analytic methods on a variety of government and commercial data sources to assess relative contractor performance risks in Air Force acquisition contracts and programs. The authors of this report discuss initial insights and approaches in exploring and prototyping these methods. Subsequent work and further development are ongoing.

Issues

- What techniques can be employed to identify contractor performance risk indicators relative to their peers?
- How can program managers (PMs) and stakeholders leverage available government and commercial data to identify these risk indicators?
- What data sources are available and useful?
- What practical concerns arise when implementing such risk assessments?

Approach

- Review corporate risk literature and existing business intelligence tools to determine how potential risks are identified and what methods exist to reveal them.
- Compare relative conditions or performance against peers as a way to highlight outliers as potential risks for subsequent manager due diligence.
- Develop a taxonomy of these relative risk measures beyond those traditionally examined in program design and engineering.
- Identify potential data sources and algorithms for these measures.
- Obtain relevant data and develop software to extract, transform, and load these data into a custom database and processing environment.
- Build a research prototype to test and refine the concept. Cross-reference data sources to associate contracts and contractors with Air Force programs and build a simple user interface to view results from both a contractor and program perspective.

Conclusions

- Cross-indexing public and sensitive databases through modern interfaces enables new risk indicators too time-consuming to discover manually.
- Statistically comparing contractor outliers relative to peers appears to be a useful way to objectively identify potential risks; this approach identifies areas for increased management attention.

- Such automated tools can help managers focus their limited resources on potential risks buried in large, diverse data and take mitigation actions based on effectiveness and program relevance.
- Identified outliers are indicators for acquisition professionals to apply their acumen, understanding of program priorities, and acceptable levels of risk to determine the relevance and magnitude of the potential risks and what actions should be taken (if any) to mitigate them.
- Some data that are important for assessing relative contractor risks are very difficult to obtain—even for Air Force officials and federally funded research and development centers, let alone support contractors.
- Further work is necessary to develop a prototype with significant critical mass of data sources and measures to test and refine this approach. User feedback on utility and design is also needed.

Opportunities

- Integrating and assessing traditional and nontraditional data sources provide indications of potential areas of concern.
- PMs and stakeholders can use this approach to identify relative risks for further due diligence, confirmation, and proactive management.
- Additional data and the inclusion of more metrics are needed to make this more robust; data availability, accessibility, and analysis are key.
- This is a research prototype and is not ready for transition to an operational system. Despite its limitations, however, this approach is more sophisticated in some ways than other available systems and may point to features or concepts that could be added to Air Force or U.S. Department of Defense systems that assess potential contractor risks.

Contents

Figures

Tables

Summary

As all acquisition professionals are aware, there are multiple kinds of risks in systems development. Therefore, the U.S. Department of Defense (DoD) has several reporting tools, such as the Defense Acquisition Visibility Environment, Project Management Resource Tools, and others. However, these typically focus on what has been occurring with respect to cost, schedule, and performance. Although very useful, these do not provide early insights into contractor performance risks. Existing systems report outcomes (e.g., schedule slips, technical shortfalls, or cost overruns) that could be manifestations of contractor performance. Providing the acquisition professional with an earlier indication of potential contractor performance issues would allow the government program manager (PM) to address them sooner and potentially mitigate risks before they become issues. In this event, faster action lowers the resource costs to the program and the U.S. government.

To provide the acquisition professional with earlier insights into contractor performance risks, our approach gathers information from government and commercial sources. These data are gathered and combined to increase the insight into how one company is performing and then that information is compared with its peers' performance. This *relative* assessment—comparing a company to its peers to identify outliers—allows PMs and executives to focus their time and perform due diligence on the results. They then use their expertise to determine whether the outliers are relevant, how significant the risk is, if at all, and develop a mitigation strategy.

A key practical consideration in implementing a relative risk-assessment system is whether data exist and are accessible. After developing a risk taxonomy that addressed a variety of potential areas of risks, we grouped them into 11 categories with potential subfactors. We also identified a range of possible data sources that could address these risk categories and subfactors. Then, based on data accessibility, augmented by our standing as a federally funded research and development center (FFRDC) participating in a data-access pilot[1] authorized under Section 235 of the fiscal year 2017 National Defense Authorization Act and implemented by the Office of the Secretary of Defense and the Department of the Air Force (DAF), we prioritized which relative risks to address first based on data access and ease of implementation to get an initial set of working measures. These are highlighted in Table S.1.

---

[1] Public Law 114-328, Section 235, 2016, was the legal foundation for the pilot, commonly referred to as the *Section 235 Pilot*. It authorized a three-year program where FFRDCs were permitted access to sensitive information defined as "confidential commercial, financial, or proprietary information, technical data, contract performance, contract performance evaluation, management, and administration data, or other privileged information owned by other DoD contractors that is exempt from public disclosure." This research project was a participant in the Section 235 Pilot.

Table S.1. Implemented Risk Measures by Category

- **Workforce**
  - Hiring (U.S. Bureau of Labor Statistics [BLS] unemployment rate)*
  - Hiring and retention (job openings)*
  - Retention (salary growth)
  - Experience in key work area (Product and Service Code [PSC])*
- **Cost (Price)**
  - Inflation in principal place of performance
  - Overhead (declining revenue)*
  - Invest to increase capacity (jump in revenue)*
- **Financial**
  - Financial metrics (overall; operational, solvency, and liquidity)*
  - Customer base: declining revenue*
  - Declining profits
- **Stability**
  - Excluded contractor*
  - Recent and pending mergers and acquisitions*
  - Lawsuits pending
  - Lawsuit losses
  - Management turnover
  - Declining stock prices
- **Supply chain**
  - Corporatewide suppliers (risks recursively applied)*
  - Contract-specific suppliers (risks recursively applied)*

- **Influence**
  - Insignificant customer (DAF, the DoD, and federal)*
- **Performance**
  - Past or current contractor performance (Defense Contract Management Agency Program Assessment Reports)*
  - Experience in key work area (PSC)*
  - Prior experience working for the DAF, DoD, or other federal agency*
- **Security**
  - Recent cyber compromises
  - Meeting new DoD cyber requirements
  - Sufficient cleared staff
  - Sufficient cleared workspace
- **Infrastructure**
  - Production stability
- **Capacity**
  - Production capacity (economic order quantity)
  - Invest to increase capacity (jump in revenue)*
- **Future**
  - Low corporate research and development (R&D) levels
  - Low government-funded R&D

- ***Cross-cutting***
  - News alerts (keyword filters)*

NOTES: Measures are unordered within and between each category, and there may be partial correlations between different measures within each category.
* = Text in purple means there is some level of implementation in the current prototype.

Once access to the data listed above is obtained, we can use the information to reveal the risk indicators in the taxonomy. Figure S.1 illustrates one example of how we can combine data from disparate sources to identify an overall relative risk in a category.

The middle of the figure shows that the Project Management Resource Tools (PMRT[2]) provides contract numbers for Air Force programs and their major contracts. For each specific contract of interest, we can then use the contract number to look up the primary work location in the Federal Procurement Data System–Next Generation (FPDS-NG). That location can be mapped to BLS data to ascertain the unemployment rate for the region containing the primary

---

[2] PMRT is a software application acquired, operated, and maintained by the Assistant Secretary of the Air Force for Acquisition, Technology, and Logistics. All DAF acquisition programs are required to report progress and status via this software.

work location.[3] Comparing the unemployment rate with other regions yields a relative risk rating for hiring (e.g., if unemployment is relatively low, then it would be generally harder to hire—thus, the relative risk is higher).

The flow on the right side of the figure illustrates how data are used for a second related hiring risk measure. Here, we take the contract number and obtain the primary work location[4] as before, but we look up the Consumer Price Index (CPI) for the area. If the CPI is higher relative to other areas, then it may be harder to attract new workers to the area, resulting in a higher relative risk. We have performed some preliminary research on this potential risk measure but have not yet implemented it.

Finally, the left side of the figure shows a third potential risk measure. Here, we would seek the actual proposal for a particular contract to see what it says about the staffing plan and any hiring needs. Such an approach would be more specific and contain deeper insight into an individual contract.

Figure S.1. Integration of Data from Multiple Sources to Enable Risk Measures

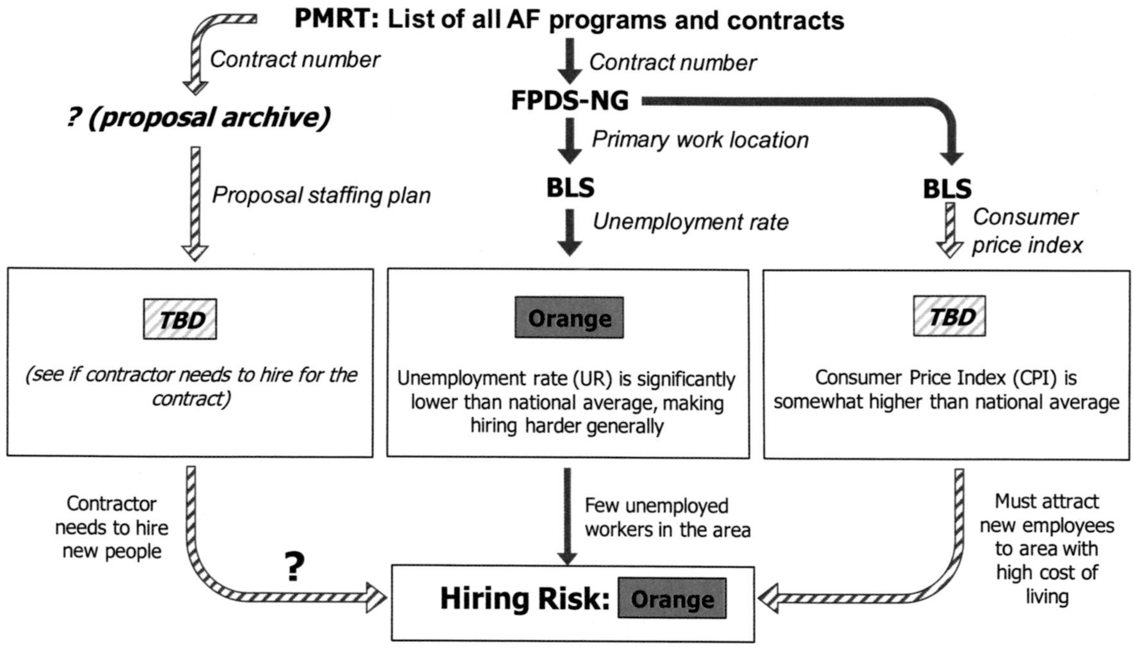

NOTE: Data elements obtained from a source are shown in italics. The thresholds for distinguishing risk levels (i.e., what constitutes "significantly" and "somewhat") are discussed in more detail in Chapter 3.

---

[3] Ideally, we would have access to all key work locations to ascertain unemployment rates beyond the primary work location, but those data are not stored in FPDS-NG, and we have not been able to access contract data on other locations.

[4] FPDS-NG only provides the primary work location. Ideally, we would use all work locations, but those data are harder to obtain (e.g., it may be buried in the proposal or contract).

These are then combined to provide an overall relative risk rating for the category of interest using algorithms we discuss in more detail in the main body of the report. However, because we are seeking relative risks compared with a company's peers, we generally used simple population statistical measures to rate the relative risks. We often calculated how many standard deviations (on the bad side) a company's relative risk is compared with the population norm. Thus, if positive (rather than negative) standard deviations are worse, anything less than a standard deviation was considered *green* (G). Values between +1 and +2 standard deviations were considered *yellow* (Y). *Orange* (O) was between +2 and +3 standard deviations, and anything above +3 was rated *red* (R). Variants of this approach include fixing these values from a baseline population or adjusting the values as the data change.

In addition to collecting, synthesizing, and analyzing data from a myriad of data sources, the results must be displayed in a clear and concise manner to be useful to acquisition professionals. Users of the relative contractor risk prototype interact with the data and results through a web interface. The prototype focuses on four primary views: an All Contractors page, numerous Single Contractor pages, an All Programs page, and numerous Single Program pages (see Figure S.2). Information on each of these pages is often linked. For example, the Single Contractor page displays information about (a) the specific contractor of focus for the page, (b) other contractors with whom this contractor has some kind of relationship, and (c) Air Force programs the contractor works on. Users can easily navigate among these pages, as shown in Figure S.2.

Figure S.2. User Interface Structure for the Prototype: Contractor and Program Views

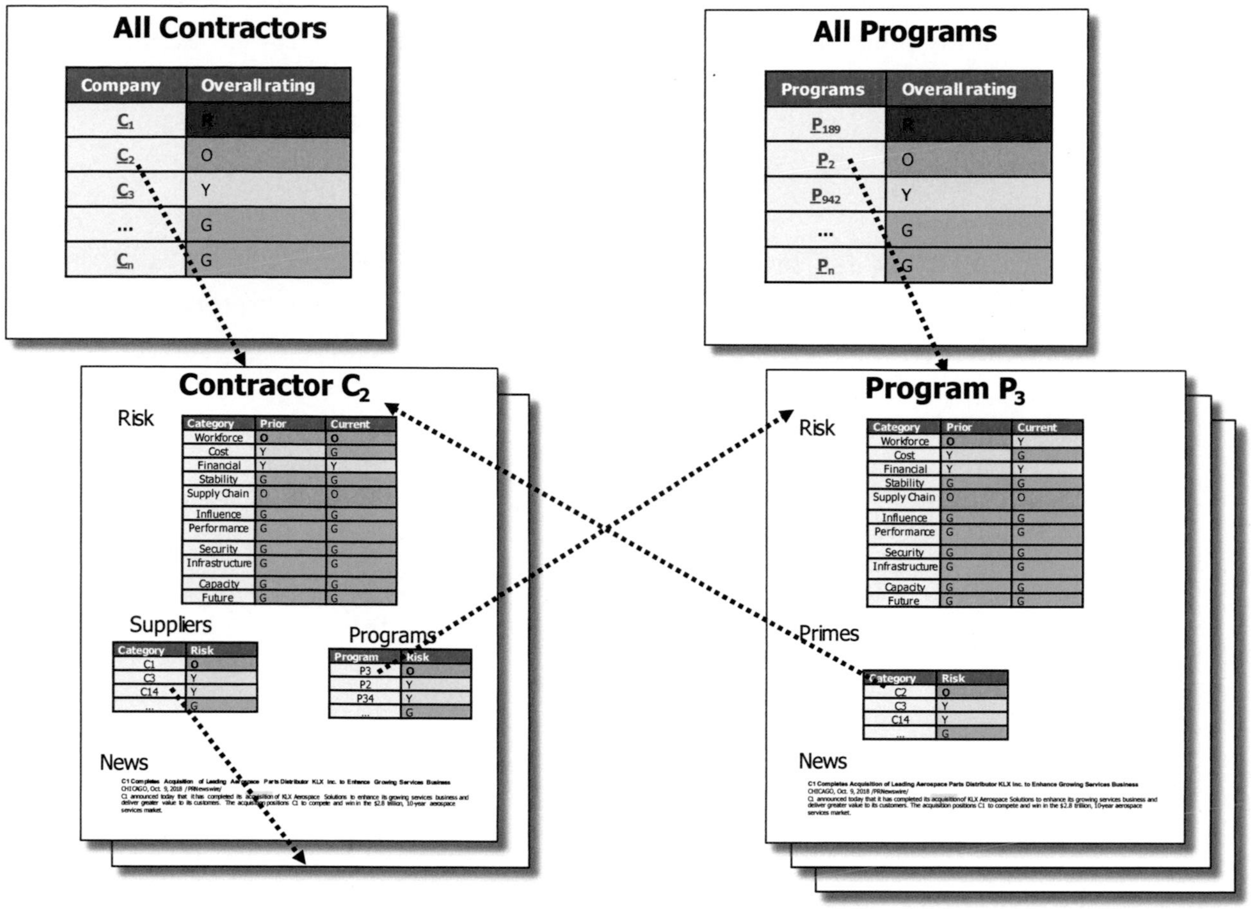

NOTE: The arrows illustrate links in the prototype between views. Thus, the Program P_3 element on the Contractor C_2 page points to the Program P_3 page. The arrow in the lower left from Supplier C_{14} points to the Contractor C_{14} page lower in the stack.

Conclusions

This approach and the initial prototype provide a visual display of *relative* contractor performance risks by collecting data from traditional and nontraditional data sources and combining them to provide indications of potential areas for subsequent attention. We use this relative outlier approach to highlight areas for additional attention.

In some sense, this approach filters outliers from large amounts of data that otherwise would be too time-consuming for managers to identify or review. Outliers are in basic areas (e.g., workforce, costs, financial health, influence, supply chains, past performance) that business theory and practice show are important to successful contractor execution. Identified outliers are risk indicators for subsequent manager attention, wherein they apply their knowledge of the situation along with further investigation to determine the magnitude of the risks and what mitigation actions should be taken (if any) based on their understanding of program priorities and level of acceptable risk to the government.

This tool does not supplant acquisition acumen; instead, it should supplement this expertise and allow professionals to focus their limited time on potential risks before they manifest into issues, thereby conserving precious program and government resources. Although other information and expert judgment are needed to ascertain whether the relative risks identified are of concern or are not being managed by the contractor, the approach is objectively applied to all companies and constitute a way to reduce the number of potential issues for further due diligence.

This remains a research prototype and is not ready for transition to an operational system. We have an initial functioning prototype, but user operation and feedback on the prototype are needed to evaluate the sensitivity and utility of the proposed measures once the prototype is matured and more data are added.

Acknowledgments

Many thanks go to our Air Force sponsor, John Miller, Senior Executive Service; and our project monitor, Mark Murphy, Senior Executive Service; and our action officer, Brian Knight, for their support, encouragement, and guidance in this research.

The approach and prototype discussed in this report builds on earlier unpublished RAND Project AIR FORCE (PAF) proof-of-concept effort in 2017 by our colleagues Chad J. R. Ohlandt, Timothy Stacey, Cole Sutera, Brian Dolan, and Stephen Joplin. Ohlandt developed the initial idea of using relative risk measures as a way to identify potential risks for further due diligence, and the remaining team explored approaches to automatically ingest and analyze unstructured public financial filings with the U.S. Securities and Exchange Commission as well as Air Force Monthly Acquisition Reports to identify important facts, cross-linkages, and situations that could form the basis for subsequent risk analysis.

We also acknowledge the valuable contributions of the other members of our project team at RAND who contributed to the design and implementation of this methodology: Edward Balkovich, Jonathan Lee Brosmer, Samantha Cohen, Lisa Colabella, Sherban Drulea, Jessica Duke, Suzanne Genc, Justin Grana, Elizabeth Hammes, Grant Johnson, Amanda Kadlec, Nirabh Koirala, Gabriel Lesnick, Edward Parker, Cole Sutera, Zoltan Szalay, Tim Webb, and David Zhang.

Reviews by Debra Knopman, Elizabeth Hastings Roer, David Loughran, and David Orletsky strengthened the clarity and quality of the report. Maria Vega provided very helpful editing. Thanks go to Obaid Younossi, Jim Powers, and Patrick Mills at RAND for their support and encouragement. Our administrative assistants—Jordan Bresnahan, Silas Dustin, Maria Falvo, and Tandrea Parrott—kept us sane and functioning.

The authors claim any errors.

Abbreviations

ACAT	Acquisition Category
AIR	Acquisition Information Repository
APB	acquisition program baseline
API	application programming interface
BI	business intelligence
BLS	U.S. Bureau of Labor Statistics
CADE	Cost Assessment Data Enterprise
CapIQ	Capital IQ
CARD	Cost Analysis Requirements Description
CMMC	Cybersecurity Maturity Model Certification
CPARS	Contractor Performance Assessment Reporting System
CPI	Consumer Price Index
DAES	Defense Acquisition Executive Summary
DAF	Department of the Air Force
DAMIR	Defense Acquisition Management Information Retrieval
DAVE	Defense Acquisition Visibility Environment
DCMA	Defense Contract Management Agency
DIB	defense industrial base
DIBNow	Defense Industrial Base Now
DoD	U.S. Department of Defense
DSS	Defense Security Service
DTIC	Defense Technical Information Center
DUNS	Data Universal Numbering System
EBITDA	earnings before interest, taxes, depreciation, and amortization
EDA	Electronic Data Access
EDGAR	Electronic Data Gathering, Analysis, and Retrieval

EOQ	economic order quantity
eTools	Electronic Tools (Defense Contract Management Agency)
EV	earned value
FAR	Federal Acquisition Regulation
FFO	funds from operations
FFRDC	federally funded research and development center
FPDS-NG	Federal Procurement Data System—Next Generation
FSRS	Federal Funding Accountability and Transparence Act Subaward Reporting System
FY	fiscal year
GAO	U.S. Government Accountability Office
IR&D	independent research and development
M&A	mergers and acquisition
MA	management assessment
MAR	(Air Force) Monthly Acquisition Report
MDA	milestone decision authority
MDAP	Major Defense Acquisition Program
NAICS	North American Industry Classification System
OSD	Office of the Secretary of Defense
PA	performance assessment
PAF	RAND Project AIR FORCE
PAR	Program Assessment Report
PDF	Portable Document Format
PM	program manager
PMRT	Project Management Resource Tools
PSC	Product and Service Code
R&D	research and development
RDT&E	research, development, test, and evaluation
RFP	request for proposal

SAF/AQX	Deputy Assistant Secretary for Acquisition Integration, Office of the Assistant Secretary of the Air Force for Acquisition Technology & Logistics
SAR	Selected Acquisition Report
SEC	U.S. Securities and Exchange Commission

1. The Challenge: Contractor Risks in Acquisition

As all acquisition professionals are aware, there are multiple kinds of risks in systems development. One such risk is poor contractor performance, which can result in program delays, increased costs, and/or reduce performance.[1] When contractors are in danger of not meeting contractual performance goals,[2] Department of the Air Force (DAF) acquisitions may not be fully aware of the shortfall until a schedule deadline is missed, government testing indicates poor performance, or costs exceed expectations. Revealing such problems later in acquisition makes it harder and costlier to rectify than if they were identified earlier. In hindsight, one can often identify the warning signs that were lost in the noise of the Air Force's acquisition enterprise. Traditionally, defense acquisition has collected structured data in the form of monthly reports; however, these systems impose a reporting burden on both the system program office and the contractor. Additionally, multiple organizations (nodes) across the Air Force acquisition community are collecting information in disparate systems. Across the Air Force enterprise, the amount of structured data can be overwhelming to acquisition leadership who want to focus their time and effort on challenges that can be effectively prevented or managed. Earned-value data[3] can provide leading indicators of problems, but timely analysis is still wanting and even earlier indicators of risk are needed.

At the same time, the availability of data continues to grow rapidly. Public sources, such as Internet news, social media, and government statistics, are now broadly available. Structured and unstructured data internal to the Air Force are increasingly available in electronic formats (e.g., Portable Document Format [PDF], briefing slides, acquisition documents). Analytic methods

---

[1] The U.S. Department of Defense (DoD) defines *risks in acquisition* as "potential future events or conditions that may have a negative effect on achieving program objectives for cost, schedule, and performance" (Deputy Assistant Secretary of Defense for Systems Engineering, 2017, p. 3). When fully understood and measured, risks have a probability and a consequence, impact, or severity of the undesired event on cost, schedule, and performance, where the potential event or condition to occur. Our approach focuses on the *risk identification* phase in the DoD's risk process planning cycle whereas subsequent due diligence of alerted relative risks would involve the *risk analysis* to determine the likelihood and consequence of the risk (Deputy Assistant Secretary of Defense for Systems Engineering, 2017). Therefore, our concept is a data-driven *risk identification methodology* focused on inclusion of data not normally employed in such identification and on prime and lower-tier contractors within acquisition programs.

[2] The focus here is on the contractor's ability to perform work. This includes financial status, level of technical expertise, management stability and experience, and other factors. The project team, in addition to a literature review, used its expertise to identify potential factors that could affect whether a company was able to complete its contractual requirements.

[3] *Earned-value management* is a program management technique for measuring performance and progress against a baseline of time-phased and budgeted work-breakdown structure. As work is performed and measured against the baseline, the corresponding budget value is "earned" (Office of the Under Secretary of Defense for Acquisition and Sustainment, 2019).

(such as text analysis and statistical correlation) have the potential to process all this information and reveal risk indicators much earlier than is possible today.

In this work, we focus on types of risk government program managers (PMs) do not typically track that relate to broader enabling factors for contractor performance (e.g., financial health of the company and workforce hiring ability) as opposed to those related to the design and technology involved in the delivered product or service or in processes for managing risks. Examples of these more-traditional approaches include DoD risk management and assessment processes (e.g., DoD Instruction 8510.01, 2017; Deputy Assistant Secretary of Defense for Systems Engineering, 2017). Fleishman-Mayer, Arena, and McMahon (2013) discuss a tool focusing on assessing system integration risks. Other approaches focus on program and contract management controls (e.g., DoD, 1988; Parker, 2011; Bounds et al., 2014; Naval Air Systems Command, 2015; Marine Corps Systems Command, 2017).

Identifying Relative Risks Through Data Integration and Analysis

Risk is a key component in any business transaction, especially transactions worth millions of dollars or more. Suppliers and buyers view risks differently, although their risks are interrelated. As a buyer, the Air Force faces risks both internal and external to the department. This research effort focuses on external risks associated with the contractors from which the Air Force acquires goods and services. Acquisition professionals should be able to inform their management of external risks by tracking business intelligence (BI) on the vendors with whom they contract. This type of BI is outward-looking and is thus a much harder problem to solve due to attaining sufficient data and managing those data over thousands of contractors. The prototype described in this report is an attempt to solve this problem by creating useful outward-looking BI for the Air Force to use in their acquisition management and oversight.

These Air Force external risks are derived from their contractors. Thus, we use various contractor-level characteristics as indicators of potential risk. These risks are outlined in Table 1.1 and include factors that can be associated with business performance of the contractor. We use existing literature, expert knowledge from experience as a project manager, and studies of acquisition oversight to identify these risks measures. In this context, a *risk* is a characteristic of an Air Force contractor that, under certain conditions, could result in poor business outcomes and thus negatively affect schedule, cost, or performance in an Air Force contract or an associated acquisition program.

Mitchell (1995) speaks to different risks that, when gone wrong, lead to negative performance of a company. These include risks related to the company's financial standing, product quality, damage to capital, reputation, schedule delays, and local social and political economic factors. Additionally, Cousins, Lamming, and Bowen (2004) highlight risks derived from social and political factors, specifically those that have to do with the outside environment (e.g., natural hazards) or environmental policy changes (e.g., physical or noise pollution limits).

Both Khan and Burnes (2007) and Ritchie and Brindley (2007) extensively review the literature on risk management in the context of suppliers. Risks that are common throughout both reviews mirror Mitchell (1995) and Cousins, Lamming, and Bowen (2004) and are characterized as the economy, the political environment, local and regional environments, natural hazards, organizational changes (e.g., regime change or new product marketing), supply chains, financial solvency, workforce, and management. Additionally, it has been found that prior work experience with the DoD—especially when successful—is an indicator of lower risk in the context of contractor performance (Bradshaw and Chang, 2013). Coupled with these academic papers, we use acquisition oversight expertise to synthesize the various indicators of risk outlined in the literature, determining which risk factors are relevant to Air Force acquisition. We also sought a more-detailed approach of revealing potential risks than the combined scoring approach from such companies as RapidRatings that integrates various financial metrics into a single risk score (RapidRatings, undated).

The approach and prototype discussed in this report builds on earlier unpublished RAND Project AIR FORCE (PAF) proof-of-concept effort in 2017 by our colleagues Chad J. R. Ohlandt, Timothy Stacey, Cole Sutera, Brian Dolan, and Stephen Joplin. Ohlandt developed the initial idea of using relative risk measures as a way to identify potential risks for further due diligence, and the remaining team explored approaches to automatically ingest and analyze unstructured public financial filings with the U.S. Securities and Exchange Commission (SEC) and the Air Force Monthly Acquisition Reports (MARs) to identify important facts, cross-linkages, and situations that could form the basis for subsequent risk analysis. Our approach is to build a taxonomy of contractor performance risks that could be measured by integrating internal Air Force and other DoD data (including proprietary and other sensitive data), public government and commercial data, and licensed commercial data. We then developed and employed analytic techniques and algorithms to combine these structured and unstructured data to identify relative contractor performance risks. The general concept is illustrated in Figure 1.1.

Figure 1.1. The Concept of Integrating Data to Measure Contractor Risks

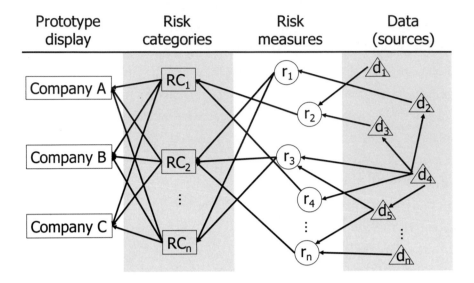

Relative (Not Absolute) Risks

At a broad level, BI is a strategy that organizations use to manage risk through the collection and analysis of data on various characteristics of a business's internal and external relations. Some organizations conduct BI in-house; however, there are many third-party organizations whose mission is to provide BI for others. These include self-service platforms, such as SAP Crystal Reports or Alteryx, that provide software with tools to organize and analyze data but leave those steps up to the company using their products (SAP, undated; Alteryx, undated). BI companies that focus on data visualization specifically include iDashboards and Dundas BI (iDashboards, undated; Dundas BI, undated). Other BI interfaces focus on data warehousing and cloud computing. These include Sisense and Oracle BI (SISENSE, undated; Oracle Solutions Business Intelligence, undated). Lastly, there are BI companies that not only provide software to manage and analyze data but use modeling and machine learning to draw insights themselves. These include such companies as InsightSquared and Domo (InsightSquared, undated; Domo, undated).

Although we acknowledge the existence of commercial BI tools, we did not find one that solves the issue of relative contractor risk management in the context of acquisition management and oversight for the Air Force or that uses statistical analysis to determine and prove which measures correlate with undesirable performance outcomes. Arguably, some of the platforms could be useful for inserting data to manage, model, and visualize. Additionally, various platforms offer risk identification and analysis through deep learning. However, in the context of acquisition management and oversight for the Air Force, they would likely neither be able to collect the needed data because of government data restrictions nor would they use methods to model the relative contractor risk for acquisition oversight managers to monitor program success. Additionally, the existing BI tools tend to be inward-looking. As previously mentioned, the Air Force as a buyer needs outward-looking BI to parse potential risks over thousands of vendors. This requires additional complexity to BI where the Air Force must collect data over a wide variety of sources, often ones that are sensitive and proprietary, to effectively proxy or capture relevant contractor risks.

A key aspect of our approach is the emphasis on *relative* risks. Here, we are comparing a company relative with its peers to identify outliers that can indicate areas where Air Force PMs and executives should focus time and due diligence to determine how real the risks are and how they can be mitigated.

This approach allows the user to compare one company against hundreds to thousands of other companies, thereby making outliers more significant. Although this approach appears potentially powerful and useful, it does have several limitations: Most of the data are available only for publicly traded companies, we are dependent on others for data collection and accuracy, and the approach does not identify program-specific risks in an absolute manner. Whether the risks exist, how significant they are, how they affect the government's level of acceptable risk, and whether actions are already being taken by the government or contractors are sufficient to mitigate them is something to be determined by the government acquisition professionals.

In an ideal world, one would test proposed metrics (such as those in this report) by seeking statistical correlation with contract outcomes. To do this would require several steps. First, access to tactical, operational data from deep within an Air Force program and the contractor's operations to get outcome data that would relate to these measures would be required; however, such data are not available outside the program or even outside the contractor's organization. Second, data that indicate whether management took action to mitigate the potential risks would be necessary to control for that variable. It is likely that there are many cases where management acted (once it found out about the situation) and other cases where it did not, resulting in very different outcomes given the same situations and thus no correlation. Lastly, other detailed variables would need to be controlled for to clarify if there is a problem (e.g., hiring risk may not be an issue if the contractor has fully staffed a contract's work). Making this determination would require controlling for numerous detailed variables (e.g., staffing plans) that are not centrally collected or available to the government. At some future point, these tactical data may be more widely available, but until then testing for correlations is probably impossible.

Also, although relative contractor risks arguably should identify potential areas of concern for program performance, our review of the literature uncovered no existing reports of tested statistical correlations of these measures to program performance. Available reports of program performance focus on case studies where the root causes of problems tend to be fairly unique. For example, root causes of critical Nunn-McCurdy cost breaches on major defense acquisition programs from 2010 to 2016 vary widely (Under Secretary of Defense for Acquisition, Technology, and Logistics, 2016, pp. 26–29; Office of Acquisition, Analytics and Policy, undated). The DoD reported trends in an inferred overall measure of program complexity and risk, but that risk measure is abstract, not specific to causes, and has too low a sample size to identify causes beyond major factors, such as work-content growth (Under Secretary of Defense for Acquisition, Technology, and Logistics, 2015, pp. 109–117). Other analyses of program performance use a case study approach that reveals the detailed differences between programs. Although useful, they reveal broader programmatic generalizations rather than specific, tactical contractor-level issues for program management attention (see, for example, Bolten et al., 2008; Porter et al., 2009; Lorell, Leonard, and Doll, 2015; Lorell, Payne, and Mehta, 2017).

We could seek anecdotes that provide examples of when a triggered risk indicator led to a contractor execution problem; these, however, are just anecdotes, not statistical tests. In the end, therefore, we rely on the business theory and practice discussed earlier in this chapter to argue that the 11 areas we list in our taxonomy relate conceptually to a contractor's execution, and we propose that integrating and assessing available data for potential risk indicators in these areas for subsequent due diligence may be a useful way to filter these data for outliers of interest.

Focusing Attention in a Sea of Data

Although these are relative risks, these indicators can help management identify areas where they can focus their time and attention. In other words, this methodology is intended to help management gain the benefits of expanding data availability by alerting to possibilities for further investigation.

As is subsequently discussed, PAF's federally funded research and development center (FFRDC) status allows us to access for this project both proprietary and Air Force–sensitive data that are necessary to model potential Air Force acquisition risk. Part of the challenge facing acquisition management and oversight in the Air Force is outward-looking data collection on contractor risk. Although the DoD tracks many data on various acquisition-related issues, McKernan et al. (2017) found that the data sources available for acquisition oversight face accessibility challenges because of the security and access restrictions necessary to control sensitive information. They assert that DoD needs to simultaneously assure access to those who need to know while protecting sensitive data. They also found that the data formats and underlying terms across these resources are inconsistent and some use outdated systems. Finally, in the context of modeling relative contractor risk, McKernan et al. (2017) found that only one of the 21 major acquisition data systems (specifically, the Contractor Performance Assessment Reporting System [CPARS]) has consistent data on contractor performance, and those data are only for past performance results, not direct indicators of future risk. As a result, DoD's data sources are not sufficient for outward-looking BI. Thus, we sought to prototype an approach that would combine DoD resources with external sources, such as commercially available BI data and other relevant data to identify a set of BI to inform acquisition risk management.

The approach also seeks a way that an automated system can scan across wide data sources for infrequent events or conditions that could have a significant effect on acquisition programs or contracts. It also seeks to exploit data sources that Air Force managers may not be aware of or have the time to check. Take, for example, bankruptcies. Such events may be infrequent and thus something that PMs do not regularly check, but if one occurs or is imminent, it would be valuable to alert the PM so that she or he can investigate and assess the magnitude of the risk and whether steps need to be taken to mitigate the risks.

How This Methodology Might Be Used

The goal of this approach is to provide acquisition officials, from PMs to milestone decision authorities, with an ability to access more data in an easy-to-understand way so they can focus their limited time on areas that require increased management attention. This approach should be useful during any phase of the acquisition process—from the pre–Materiel Development Decision phase through disposal.[4] However, it is not a replacement for risk management and must be supplemented by the user's acquisition expertise, program knowledge, and priorities (both internal and external) to appropriately highlight areas for additional focus.

Because the usefulness of the measures relies on their subsequent application by PMs and oversight managers and their expertise rather than on absolute and verified risk correlations, the usefulness of various measures may be situation dependent and lessons on their usefulness may change over time. Therefore, as experience is gained in using these measures, the users may want an ability to adjust the weights for each measure to reflect each situation or this learning.

Use by a Program Manager

Defense acquisition PMs are typically focused on executing their specific program and meeting the cost, schedule, and performance goals set and agreed to in the acquisition program baseline (APB), sometimes referred to as the contract between the PM and the milestone decision authority (MDA). Although the APB specifies the major goals for the program, it does not explicitly discuss the many underlying factors that contribute to meeting these objectives. Therefore, the PM can become singularly focused on these areas and, unbeknownst to him or her, miss other indicators of risk until they manifest as issues.

Because this methodology uses data from a variety of government-controlled, public, and commercial sources[5] and identifies relative risks compared with peers, it reveals areas where the PM should provide additional due diligence to understand the risks and whether the contractor is

---

[4] If this type of approach is going to be used in a source selection, the government should consider whether to identify this in Section M of the associated request for proposal (RFP) when factors for evaluation are identified. This approach could supplement the past performance information and any risk assessment

[5] For example, this methodology accesses data from the U.S. Bureau of Labor Statistics (BLS), Defense Industrial Base Now (DIBNow), Federal Procurement Data System-Next Generation (FPDS-NG), Federal Funding Accountability and Transparence Act Subaward Reporting System (FSRS), SEC filings, news reporting, S&P Global Market Intelligence, and more-traditional U.S. Air Force program management information systems (e.g., the Project Management Resource Tools [PMRT]).

mitigating those risks. This approach breaks down the traditional areas of cost, schedule, and performance risk into 11 categories which are further decomposed into factors as shown in Table 1.1.

Table 1.1. Components of Risk Factors to Be Considered

Workforce	Cost (Price)	Financial	Corporate Stability
Cleared staff, meeting staffing plan, attractiveness of work location, inflation at work location, prior federal work, experience (leadership and staff), research and development (R&D) spending to support design teams, retention (salary growth)	Inflation in primary work location(s), large revenue increases (does capacity need to increase?), decreases in revenue (increases in overhead), production costs, inflation (materiel)	Credit Health Score (solvency, liquidity, operational), declining profits, solidity of customer base	Recent/pending mergers and acquisitions (M&As), declining stock price, C-Suite turnover, lawsuits (pending or losses)

Supply Chain	Influence	Contractor Performance	Security
Suppliers (corporate-wide, contract-specific), reduced availability of sources, quality assurance, parts (tampering, counterfeit)	Significant customer (Air Force, DoD, federal)	Prior work experience (specific product, service, or technology), federal, DoD, commercial reputation	Recent cyber compromises, meeting cyber requirements, sufficient cleared staff, sufficient cleared workspace

Infrastructure	Capacity	Future
Production stability, new infrastructure required	Production capacity (economic order quantities [EOQs]), large increases in revenue (more needed capacity)	Low R&D investments (corporate, government-funded), published technology works, patent applications

For example, the PM may not know that the contractor is having difficulty filling critical personnel positions until he or she sees a degradation in earned value (EV) metrics (e.g., the schedule slips because work packages on the critical path are not completed on time). If sufficient types of data were available, this approach conceptually could indicate that the Workforce category is rated worse than good, allowing the PM to drill down and see that there are several vacancies at the work location that the contractor is trying to fill. Or that the contractor received a larger contract from another government entity that requires the same kinds of skills. This could indicate to the PM that his or her contractor team could be redirected to a program of greater importance to the contractor. Knowing these are possibilities supported by data, the PM could address these risks with the contractor before they manifest as issues.

This approach could be even more useful for an oversight manager who is responsible for a portfolio of programs. When overseeing multiple programs, the manager is very dependent on the reporting of the various programs. Although this approach does not identify absolute risks, it does use data sources not traditionally used by acquisition senior leaders and combines them to highlight potential areas for increased management attention prior to risks manifesting themselves as issues. This approach supplements that traditional reporting (e.g., MARs, selected acquisition reports [SARs], Defense Acquisition Executive Summary [DAES]), with additional data allowing the oversight manager to quickly assess the relative contractor performance risks across multiple programs (or contractors) via a dashboard. Her or she will see the categories in Table 1.1 for each contractor of interest and see how many Air Force programs this contractor participates in. This would rapidly give the oversight manager the scale of a risk and whether it affects parts of his or her portfolio. Armed with these indications, additional management attention can be applied to the individual programs, if deemed necessary, or to the contractor's corporate leadership as appropriate.

As with the PM, this approach supplements other available information and needs to be reviewed within current priorities, both internal and external to the portfolio. This type of insight will help focus the acquisition executive on the more-pressing issues of the day.

Prototyping to Test This Approach

This is the final report for Phase I of an effort to test the approach outlined earlier by building a prototype that uses actual data to calculate relative risk measures and present them to users to assess its value and refine the approach. This research prototype is not ready for transition to an operational system. Such operationalization would require a significant investment to be determined. Although we have an initial working functioning prototype, further work is needed to evaluate the sensitivity and utility of the proposed measures once the prototype is matured and more data are added. Further work is necessary to develop a prototype with a significant critical mass of data sources and measures to test and refine this approach. User feedback on utility and design is also needed. Despite its own limitations, however, this approach is more sophisticated in some ways than other available systems and may point to features or concepts that could be added to Air Force or DoD systems that assess potential contractor risks.

Organization of the Report

Chapter 2 presents a taxonomy of contractor risks and identifies potential data sources that could be used to measure those risks.

Chapter 3 describes the algorithms developed for individual risk measures for the implementation prototype. It also describes how these relative risk measures are combined to obtain summary risk scores and ratings for the user.

Chapter 4 describes the architecture and approach taken to prototype this relative risk methodology using actual data.

Chapter 5 summarizes the insights and lessons learned along with suggested next steps for implementing relative risk measures using these expanding data sets.

2. Taxonomy of Relative Risks

A taxonomy of risks and potential data sources was developed to begin laying out the space of possible relative contractor risk measures that could be implemented and to help prioritize them based on estimated ease of implementation.

Categories of Contractor Risk

Using what we learned from our literature review and team experience, we developed a list of possible risk measures, grouping them into the 11 categories (plus cross-cutting measures that involve more than one category) as discussed in Table 1.1 of Chapter 1.[1] Note that we are focusing in this work on types of risk related to enabling factors for contractor performance as opposed to those related to the design and technology involved in the delivered product or service. These enabling factors are, perhaps, those that receive the least attention and are hard to characterize because they rely on various data sources that are hard to access and for which integration and analysis are required. Hence, that is our focus for applying data access and analytic techniques in this project.

Identified risk measures were developed from expert knowledge and experience in Air Force acquisition, a literature review focusing on risk in the context of business intelligence outlined above, and data availability. In some cases, we know that data exist but are difficult or practically impossible to access at present. Here, we erred on the side of including risk measures that need such data to seek a more inclusive set of the possible.

Potential Data Sources

Key practical considerations in implementing a risk-assessment system are whether data exist and are accessible. We identified a range of possible data sources that, when combined, could give insight into relative contractor risks. Table 2.1 lists these data sources, along with our rough assessment of their accessibility to us as an FFRDC under the Section 235 pilot.[2] Here the easiest

---

[1] The categorization is based on logical relationships between the measures. We did not examine what the effect of alternative groupings might be, but note that the functions described in later chapters that summarize ratings are constructed in such a way as to ensure that abnormal scores are not lost regardless of grouping.

[2] Public Law 114-328, Section 235 (2016) was the legal foundation for a pilot sponsored by the Office of the Secretary of Defense (OSD), commonly referred to as the Section 235 Pilot. It authorized a three-year program in which FFRDCs were permitted access to *sensitive information* defined as "confidential commercial, financial, or proprietary information, technical data, contract performance, contract performance evaluation, management, and administration data, or other privileged information owned by other contractors of the DoD that is exempt from public disclosure." This research project was a participant in the Section 235 Pilot.

to access are given a rating of 4, while the hardest are rated as 0. These ratings may change not only for us (as an FFRDC) with effort but also for others who may implement such a risk assessment system.

Although this is only a partial list of available data sources, its length helps to illustrate that too many data sources and too much information are available for PMs and oversight officials to consume. That motivation helped lead to the current methodology to identify out of all that data which contractors have measures that are well outside the norm for the population, thus highlighting and alerting possible risks for due diligence by the PM or oversight officials.

The second aspect of the table is Ease of Access score. This score is purely a practical indicator for our prototyping effort, not a measure of the relevance or importance of a particular data source. If, for example, a certain measure and its associated data are deemed of high value to the Air Force but access is difficult, then the Air Force would need to seek a remedy to improve access either for our prototype or for the department's own implementation of such a system.

Table 2.1. Possible Data Sources and Ease of Access for This Effort

Data Source	Ease of Access
Acquisition Information Repository (AIR)	4 *(easiest)*
American Community Surveys (U.S. Census Bureau)	4
BLS	4
Cost Assessment Data Enterprise (CADE)	4
Capital IQ (CapIQ)	4
Corporate websites	4
Defense Acquisition Management Information Retrieval (DAMIR)	4
Defense Acquisition Visibility Environment (DAVE)	4
DIBNow	4
EV Management Central Repository	4
FSRS (subcontracts > $25,000)	4
FPDS-NG	4
Bid protests to the U.S. Government Accountability Office (GAO)	4
Government-Industry Data Exchange Program	4
Job postings *(e.g., CareerBuilder, ClearanceJobs, Dice, Glassdoor, Indeed, Job.com, LinkedIn, Monster)*	4
LexisNexis	4
News	4
Patents database	4
Press releases	4
PMRT	4
RFPs (FedBizOpps.gov)	4
SEC's Electronic Data Gathering, Analysis, and Retrieval (EDGAR)	4
SEC filings	4
System for Award Management	4
Web of Science	4
CPARS	2
Court filings	2
Cybersecurity regulatory compliance	2
Defense Contract Management Agency (DCMA) Program Assessment Reports (PARs)	2
Contractor's Proposal	1.5
DMCA Electronic Tools (eTools)	1.5
Electronic Data Access (EDA)	1.5
Defense Security Service (DSS)	1
Defense Technical Information Center (DTIC)	1
Air Force Total Ownership Costs (Visibility and Management of Operating and Support Cost) (Operating and Support Actuals)	0 *(hardest)*
Air Force Cost Analysis Agency-Service Cost Position	0
Bills of lading (ports)	0
Committee on Foreign Investment in the United States	0
Contracting Officer Representative Tracking tool	0
Defense industrial base (DIB) cybersecurity filings	0

NOTE: Ease of access is our approximate assessment of how hard it is for FFRDCs under the Section 235 pilot to gain access. Values may be adjusted with additional effort. Higher numbers are easier than lower numbers. Values can be 4, 2, 1.5, 1, or 0 and were set in this nonlinear fashion so that the easiest sources to access rate higher than in a linear rating. Job posting website examples are listed alphabetically and are not prioritized.

Product and Service Codes

Product and Service Codes (PSCs) are used by the government to group products, services, and other activities for contract reporting. As the U.S. General Services Administration indicates:

The Product and Service Codes (PSCs) . . . describe products, services, and research and development (R&D) purchased by the federal government. These codes indicate "WHAT" was bought for each contract action reported in the Federal Procurement Data System (FPDS). . . . In many cases, a given contract/task order/purchase order will include more than one product and/or service. In such cases, the "Product or Service Code" data element code should be selected based on the predominant product or service that is being purchased (U.S. General Services Administration, 2015, pp. 5–6).

Thus, the readily available PSCs in FPDS-NG provide a way to easily identify the main type of work performed for each federal contract, including all Air Force and other DoD contracts whose existence or amount is not classified.

PSCs are four-character codes. If the first character is a numeral, then the work is to deliver a *product*. If the first character is a letter, then the work is a *service*. Coding is provided in a hierarchical fashion so that groups of similar work share leading characters and subsequent characters in the code provide further specificity on the type of work.

We use PSCs in the prototype to identify similar types of work without trying to be so specific that minor differences indicate fundamentally different types of work. For example, PSC category AC11 codes for *R&D-Defense System: Aircraft (Basic Research)*, whereas category AC12 codes for *R&D-Defense System: Aircraft (Applied Research/Exploratory Development)* (U.S. General Services Administration, 2015, p. 31). In our application, the distinction at the fourth character level is usually not significant when trying to assess things, such as workforce skills and discipline, so we would disregard the fourth character. The question, then, is how many characters to use when comparing PSCs so that we group similar types of work without being too specific. To achieve a balance, we developed the usage scheme shown in Table 2.2 for the prototype.

Table 2.2. PSC Precision Approach for the Prototype

First Character	Number of Leading Characters Used	Reasoning
A	3	PSCs that start with an "A" are for R&D. The distinctions at the third-character level are important. For example, category AC1 codes for *R&D-Defense System: Aircraft*, whereas AC2 codes for *R&D- Defense System: Missile/Space Systems* (U.S. General Services Administration, 2015, p. 25)—an important distinction.
B–Z	2	Other service codes were used at the two-character level. This distinguishes between areas, such as *Special Studies/Analysis* (category B5) from *Information Technology and Telecommunications* (category D3) and *Quality Control* (category H2) (U.S. General Services Administration, 2015, p. 61). This may be sufficient generally, but, in some cases, we might want to employ further distinctions in the future (e.g., between *Quality Control—Guided Missiles* [category H114] from *Quality Control—Aircraft and Airframe Structural Components* [category H114]).
Numeric	2	This level of specificity provides a distinction between, for example, *Guided Missiles* (group 14) and *Aerospace Craft and Structural Components* (group 15).

Experimentation with the data and adjustment for increasing or decreasing the rates of risks indicated will help dictate what level of specificity is best. This may be determined over time as a system default or as a user-specified parameter that can be adjusted for different runs of the system.

Using Multiple Data Sources to Measure Risks

Once access to the data sources listed in Table 2.1 is obtained, we can use the information across multiple sources to estimate the relative risks in the taxonomy. Recall from Figure 1.1 that, for each relative risk measure, one or more data sources are used to calculate the measure.

Figure 2.1 illustrates how using different data sources make this possible. In this example, the middle of the figure shows that PMRT provides contract numbers for Air Force programs and their major contracts. For each specific contract of interest, we can then use the contract number to look up the primary work location in FPDS-NG (see Table 2.3). That location can be mapped to BLS data to ascertain the unemployment rate for the region containing the primary work location. Based on how that unemployment rate compares with other regions for other contractors should yield a relative risk rating for hiring (e.g., if unemployment is relatively low, then it would be generally harder to hire—thus, the relative risk is higher).[3] This risk measure is discussed further in the next chapter and was implemented in the prototype.

---

[3] Further analysis to test this relationship could be valuable for future research.

Table 2.3. Selected Data Available from FPDS-NG

Data Element	Example
Effective date	12/12/2016
Estimated ultimate completion date	01/04/2017
Procurement Instrument Identifier (e.g., contract number)	FA877216F1026
Modification number	P00001
Type of contract	Firm fixed price
Action obligation amount: Current ($)	$146,250.00
Action obligation amount: Total ($)	$1,990,633.88
Funding agency identifier	5,700
Funding agency name	DAF
Vendor name	Lockheed Martin Corporation
Data Universal Numbering System (DUNS) number	147460526
Number of employees	140,000
Annual revenue	$41,899,999,232
Principal place of performance: city	King of Prussia
Principal place of performance: county	Montgomery
Principal place of performance: state	Pennsylvania
Principal place of performance: zip code (+4):	19406-2902
PSC	D306 (information technology and telecom systems analysis)
Principal North American Industry Classification System (NAICS) code:	541512
DoD Acquisition Program	000 [not provided]

SOURCE: Authors' analysis of 2017 data from U.S. General Services Administration.

The flow on the right side of Figure 2.1 illustrates the data usage for a second, related hiring risk measure. Here we take the contract number and obtain the primary work location in Table 2.3, but we look up the Consumer Price Index (CPI) for the area. If the CPI is higher relative to other areas, then it may be harder to attract new workers to the area, resulting in a higher relative risk. We have performed some preliminary research on this potential risk measure but have not yet implemented it.

Finally, the left side of the figure shows a third potential risk measure. Here, we seek the actual proposal for a particular contract to see what it says about the staffing plan and any hiring needs. Such an approach would be more specific and contain deeper insight into an individual contract, but we do not know of a readily available archive for Air Force proposals.

Chapter 3 discusses in detail the implemented risk measures and associated data sources used. Other possible risk measures and sources are discussed in Appendix A, and additional research may uncover new data sources with potential utility. For example, DCMA staff or databases may have insights into staffing issues at major contractor sites.

Figure 2.1. Integration of Data from Multiple Sources to Enable Risk Measures

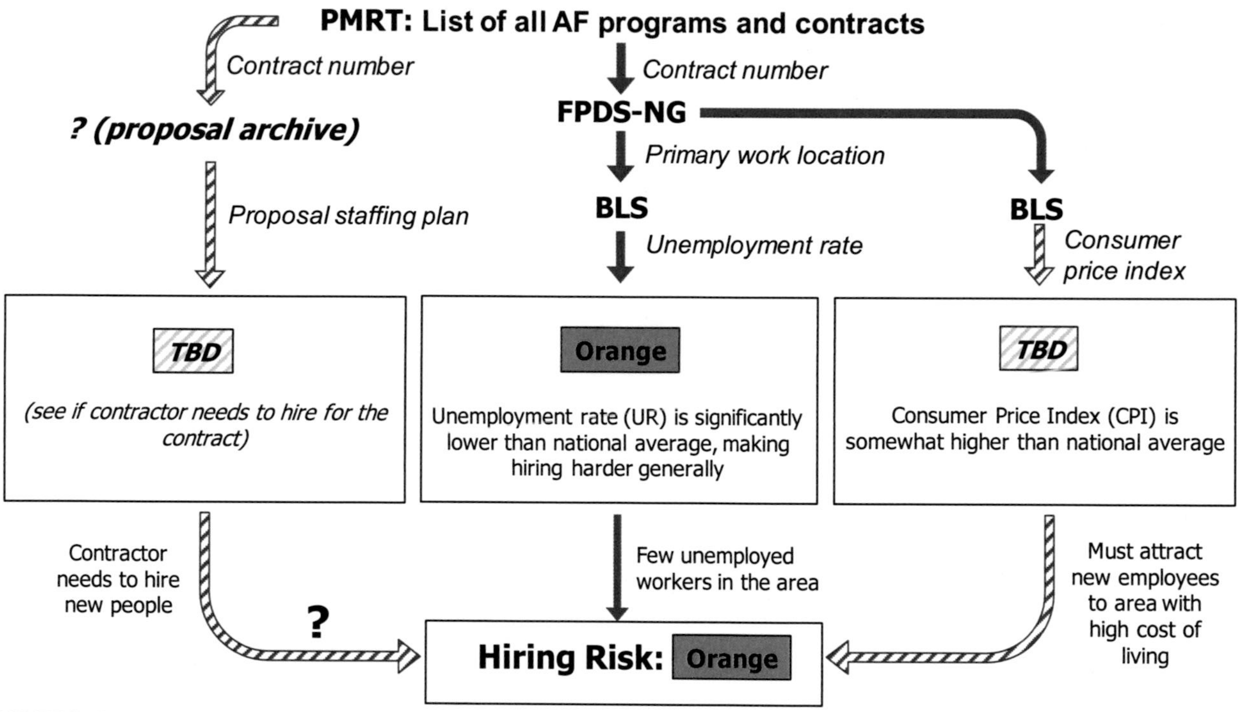

NOTES: Data elements obtained from a source are shown in italics. The thresholds for distinguishing risk levels (i.e., what constitutes "significantly" and "somewhat") are discussed in detail in Chapter 3.

Prioritizing Risk Measures for Implementation

Because we identified a large number of potential relative risk measures to implement in the prototype, we developed the simple prioritization scheme shown in Table 2.4 to select the ones to implement first. Here, we simply multiply the scores indicated earlier for data access ease and relative risk measure implementation difficulty. Note that we put more emphasis on the data access challenges in that more difficult ones have lower numbers, resulting in lower prioritizations. This scheme does not reflect which measures are more important—just which are easier to implement as a first pass at trying to get multiple measures into the initial version of the system to begin getting a sense of how the system might operate. Within those that are easiest to implement (those with scores of 4 x 4 = 16), we sought sponsor feedback for the ones in which they have the most interest. In the long run, there may be high-priority measures for the Air Force that use data that are hard to obtain or whose risk measures are difficult to implement; those can be prioritized by the Air Force in later phases of the effort as resources allow.

Table 2.4. Prioritization Scheme for Implementing Risk Measures

			Relative Risk Measure Implementation Difficulty					
			Low		Medium		High	
		Scores	4		3		2	
	Green	4	16	Green-Low	12	Green-Medium	8	Green-High
Data	Yellow	2	8	Yellow-Low	6	Yellow-Medium	4	Yellow-High
Access	Pink	1.5	6	Pink-Low	4.5	Pink-Medium	3	Pink-High
Ease	Red	1	4	Red-Low	3	Red-Medium	2	Red-High
	Black	0	0	Black-Low	0	Black-Medium	0	Black-High

NOTE: The prioritized score is the product of the difficulty of relative risk measure implementation and the ease of data access score.

Using this prioritization scheme, we focused our initial prototyping efforts on the relative risk measures listed in Table 2.5. Here we list the most promising measured in each risk category. Unfortunately, some of those risks were very challenging compared with those in other categories and were not implemented in the current version. For example, none of the relative risks in the Security category were implemented. Note that we currently do not have risk measures implemented in the prototype for three categories: Security, Infrastructure, and Future.

Table 2.5. Implemented Risk Measures by Category

- **Workforce**
  - Hiring (BLS unemployment rate)*
  - Hiring and retention (job openings)*
  - Retention (salary growth)
  - Experience in key work area (PSC)*
- **Cost (Price)**
  - Inflation in principal place of performance
  - Overhead (declining revenue)*
  - Invest to increase capacity (jump in revenue)*
- **Financial**
  - Financial metrics (overall; operational, solvency, and liquidity)*
  - Customer base: declining revenue*
  - Declining profits
- **Stability**
  - Excluded contractor*
  - Recent and pending M&As*
  - Lawsuits pending
  - Lawsuit losses
  - Management turnover
  - Declining stock prices
- **Supply chain**
  - Corporatewide suppliers (risks recursively applied)*
  - Contract-specific suppliers (risks recursively applied)*

- **Influence**
  - Insignificant customer (Air Force, DoD, and federal)*
- **Performance**
  - Past and current contractor performance (DCMA PARs)*
  - Experience in key work area (PSC)*
  - Prior experience working for Air Force, DoD, and federal*
- **Security**
  - Recent cyber compromises
  - Meeting new DoD cyber requirements
  - Sufficient cleared staff
  - Sufficient cleared workspace
- **Infrastructure**
  - Production stability
- **Capacity**
  - Production capacity EOQ
  - Invest to increase capacity (jump in revenue)*
- **Future**
  - Low corporate R&D levels
  - Low government-funded R&D

- ***Cross-cutting***
  - News alerts (keyword filters)*

NOTES: Measures are unordered within and between each category, and there may be partial correlations between different measures within each category.
* = Text in purple means there is some level of implementation in the current prototype.

19

3. Relative Risk Measures and Equations

For each measure that we implemented in the prototype, we developed an algorithm for calculating relative risk. This chapter describes those algorithms for risks grouped by their category. Note that we currently do not have risk measures implemented in the prototype for three categories: Security, Infrastructure, and Future.

Using Statistics to Measure Relative Risks

We are seeking *relative* risk measures compared with a company's peers. Therefore, we use simple population statistical measures to rate the relative risks. One useful measure of how far a company is from the average is the standard deviation.[1] In these cases, anything less than +1 standard deviation was considered *green* (G). Values between +1 and +2 standard deviations were considered *yellow* (Y). *Orange* (O) was between +2 and +3 standard deviations, and anything above +3 was rated *red* (R).

Data Limitations

Note that the data sources listed in Table 2.1 do not provide universal coverage across all companies. For example, financial data from SEC filings are only for publicly traded companies. Other sources (such as CapIQ, Factset, and Eikon) obtain financial data on some private companies, but we do not expect that to be comprehensive. The same can be said for other data sources (e.g., GAO's bid protest data do not include protests filed in the Court of Federal Claims). Although this does lead to a lack of coverage when data are not available (i.e., cases of not identifying issues from lack of data), it could potentially bias the population against which we use statistical measures to identify outliers. For example, privately held companies might perform worse generally than publicly traded companies in certain financial measures. Because these data are not available, we do not know. Our approach, however, should be relatively robust against the available population so that the user will be alerted to outliers for further analysis,

---

[1] We do not assume any population distribution but note the following to get some perspective on how much of the population may be more than two or three standard deviations from the average (norm) on the high side. If, for example, the population happened to have a normal (Gaussian) distribution and the sample size was high, then we would expect (as the sample size increased) that we would have about 2.1 percent of the population in Orange (between two and three standard deviations on the high side) and about 0.1 percent in Red (more than three standard deviations on the high side). If, in the extreme case, we do not know what the distribution is, then Chebyshev's inequality tells us that, regardless of the population distribution, at least 75 percent of the population is within two standard deviations of the mean and at least 88.89 percent is within three standard deviations of the mean (Chernick, 2011). We could, of course, count for each measure the percentage of companies in each rating. Such a count may be a useful addition to future versions of the prototype so that the user has a sense of how many companies are being raised by the system for consideration.

assuming the population size is significant. The idea in our approach is to flag the worst out of the population we have. If all of that population performs poorly against a larger population for which we do not have data, at least we are able to say which are the worst among that set.

Workforce Category

Hiring (U.S. Bureau of Labor Statistics Unemployment Rate)

Motivation. Here we check whether unemployment is relatively low in the principal place of performance. If so, then hiring should be more difficult generally in the area,[2] making it harder for companies to increase capacity or meet new workforce skill demands as they grow or shift to new types of work.

Approach. For each DoD contract, the system uses the reported principal place of performance in FPDS-NG. The system then looks up the BLS unemployment rate for the reported region in which the principal place of performance is located. Relative contractor risks are then rated based on whether the unemployment rate is more than a standard deviation above the U.S. population. Ratings range from yellow (if the contractor is only one standard deviation worse than average) to red (if the contractor is more than three standard deviations worse than average).

Improvement options. Ideally, we would compare job-specific unemployment rates in the area with the primary PSC for the contract work (or better yet, contractor shortfalls by job). These improvements may be considerations for future upgrades (if appropriate data can be accessed).

Hiring (Job Postings)

Motivation. Here we wanted to see whether job postings could be used to indicate hiring needs for each contractor. Job postings should give some indication of internal workforce needs by discipline and work location.

Approach. We purchased about 405,000 job postings from Monster.com[3] in two batches approximately three months apart to obtain a working set of actual data focused on engineering, production, and software positions (see Table 3.1). FPDS-NG was used to determine the

---

[2] Note that the precision available for areas depends on the precision in the BLS data. Although greater precisions might be better, it does provide a general sense by region, and there is more employment flexibility within a region than moving to a new job in a different region. These areas are based on the BLS defined *area code*, a five-digit number typically referencing either a metropolitan area or county. For a full discussion, see BLS, 2019.

[3] As mentioned in Table 2.1, there are many sources for online job postings. We are using Monster.com data as a proof of concept, not as an advocacy of a particular product. We also evaluated data from other similar websites, such as Indeed.com and ClearanceJobs.com. If budget were not a concern, including data from all of these sites plus others would yield the best result. Alternatively, the product could switch from one provider of job data to another (by rebidding the contract periodically), and the risk calculation would only need minor changes to the specific data ingestion code.

principal place of performance for each contract. The system then looks for open positions with known defense contractors (companies identified by other data sources ingested by the relative contractor risk prototype). Each of these job openings lists (1) the amount of time that this position has been open and (2) a classification code describing the position. The system then compares the length of time that the position for the defense contractor has been open with the average length of time that positions of this type remain open around the country. If the job has been open for significantly longer than other jobs in this category, then the contractor is identified as having a relatively higher risk that it will be unable to staff its workforce.

Improvement options. Other more sophisticated risk measures are possible. For example, FPDS already identifies the primary PSC associated with any individual contract. If a mapping existed between these PSCs and the types of jobs typically required to perform work of that type, then a risk condition could be created that evaluates how difficult it is to hire workers in those job categories at the planned place-of-performance against national averages. Additionally, integrating additional data sources, either additional job-posting sites (such as ClearanceJobs.com) or external job postings of defense contractors themselves, could provide additional data points about the overall risk that contractors face in staffing their workforce.

Table 3.1. Monster.com Job Categories in the Sample Set of Jobs Postings

Monster.com Code	Job Category
13-1111.00	Management analysts
15-1111.00	Computer and information research scientists
15-1121.00	Computer systems analysts
15-1131.00	Computer programmers
15-1132.00	Software developers, applications
15-1133.00	Software developers, systems software
15-1143.00	Computer network architects
15-1152.00	Computer network support specialists
15-1199.00	Computer occupations, all other
15-1199.01	Software quality assurance engineers and testers
15-1199.02	Computer systems engineers/architects
17-2011.00	Aerospace engineers
17-2041.00	Chemical engineers
17-2061.00	Computer hardware engineers
17-2071.00	Electrical engineers
17-2072.00	Electronics engineers, except computer
17-2111.00	health and safety engineers, except mining safety engineers and inspectors
17-2111.01	Industrial safety and health engineers
17-2111.02	Fire-prevention and protection engineers
17-2112.00	Industrial engineers
17-2112.01	Human factors engineers and ergonomists
17-2131.00	Materials engineers
17-2141.00	Mechanical engineers
17-2199.00	Engineers, all other
17-2199.04	Manufacturing engineers
17-3012.02	Electrical drafters
17-3013.00	Mechanical drafters
17-3021.00	Aerospace engineering and operations technicians
17-3023.01	Electronics engineering technicians
17-3023.03	Electrical engineering technicians
17-3024.00	Electro-mechanical technicians
17-3026.00	Industrial engineering technicians
17-3027.00	Mechanical engineering technicians
17-3029.05	Industrial engineering technologists
17-3029.06	Manufacturing engineering technologists
17-3029.09	Manufacturing production technicians
51-2011.00	Aircraft structure, surfaces, rigging, and systems assemblers
51-2022.00	Electrical and electronic equipment assemblers
51-2023.00	Electromechanical equipment assemblers
51-2099.00	Assemblers and fabricators, all other
51-4011.00	Computer-controlled machine tool operators, metal and plastic
51-4012.00	Computer numerically controlled machine tool programmers, metal and plastic
51-4041.00	Machinists
51-4061.00	Model makers, metal and plastic
51-4081.00	Multiple machine tool setters, operators, and tenders, metal and plastic
51-4121.06	Welders, cutters, and welder fitters
51-4199.00	Metal workers and plastic workers, all other
51-9041.00	Extruding, forming, pressing, and compacting machine setters, operators, and tenders
51-9195.07	Molding and casting workers
51-9199.00	Production workers, all other

Experience in Key Work Areas (Prior Product and Service Code Experience)

Motivation. FPDS-NG indicates the main type of work for each contract by providing a PSC. Because FPDS-NG provides a complete history of all federal contracting, we should be able to see whether and how much prior work each company has done in a particular area (assuming we can correlate companies over time given M&As, reorganizations, divestitures, and other factors that make it difficult to track companies and their workforce). Of course, this only includes federal work, but such a measure should help indicate whether the PM should further examine a company's experience (e.g., during source selection or in the early stages of the contract).

Approach. First, the system examines all prior contracts for the past few years (we downloaded four years of DoD contract data from FPDS-NG for the prototype covering 100,766 contracts) to build a history of prior work (number of contracts and obligation dollars) by PSC. The PSC resolution is shown earlier in Table 2.2.

We then look at the number of contracts that each contractor has been awarded by PSC. In our current implementation, if the contract in question is for at least $500,000 (which is twice the nominal simplified acquisition threshold[4]) and the contractor has not been awarded another contract in the PSC grouping in last two years (fiscal year [FY] 2017 or 2018), then the contractor is considered to have a yellow-level risk to alert to the PM. Otherwise, the contract is green.

Improvement options. Ideas for improving the fidelity or scope of measuring past work experience include the following:

- expanding past contracts to include their period of performance for the past two years (not just the initial award date)
- setting the rating thresholds based on the population distribution in PSC obligation concentration
- further adjusting the PSC resolution in particular PSCs (e.g., to distinguish between areas of quality control)
- comparing the amount of prior work experience in dollars to the company's revenue
- comparing the current contract's dollar size to the dollar value of the past total three-year annual obligations in this PSC
- seeing how long it has been since any work has been obligated to the company in this PSC.

---

[4] DoD (2018) increased the normal simplified acquisition threshold to $250,000.

Cost Category

Overhead: Declining Revenue (Price Risk: Overhead)

Motivation. If a company is experiencing a declining trend in total revenue, then there is a risk that costs (and prices paid by the government) may increase if overhead costs are not reduced commensurate with the revenue declines. Even if overhead is reduced, there are often fixed costs that cannot be effectively reduced at the same rate of the revenue declines (i.e., the reverse of efficiencies of scale because of increased production or integration from mergers or reorganizations).

Approach. For each company, we used annual total revenue $R(t)$ for year t. Public U.S. companies report total revenue in their quarterly and annual SEC filings. These data can be obtained from public postings in the SEC's EDGAR system or purchased in a structured format from various companies[5] that collect and structure corporate financial data (including international and some private companies). The algorithm is as follows:

1. Extract annual total revenue $R(t)$ from the data source (e.g., corporate income statements). (Note: Revenue reflects delivered products and progress on other contracts. If a company goes through an M&A, then it will restate/reclassify revenue so that it can be compared across years [e.g., by adding or subtracting the revenue from a unit that is bought or sold, respectively][6]).

2. Calculate the three-year moving average (R_{ma}) of total annual revenue.

$$R_{ma}(t) = (R(t-2) + R(t-1) + R(t)) / 3$$

(Note: We may decide later that a two-year moving average is needed if this smooths out the data too much.)

3. Calculate the percentage change R_c in moving average from two years ago.

$$R_c(t) = (R_{ma}(t) - R_{ma}(t-2)) / R_{ma}(t-2)$$

---

[5] Some commercial data systems that we identified include CapIQ (S&P Global), Eikon (Refinitiv), and FactSet.

[6] We do not want to use "bookings" or "orders" because they are more like obligations from the DoD. They go into backlog but do not reflect progress made. If we had looked at backlog, it would probably be split between funded and unfunded (i.e., the latter would be contract options not yet executed).

(Note: We may decide later to only compare with last year's moving average if the moving average from two years prior smooths out the data too much.)

4. Rate the *declining revenue* risk using the thresholds in Table 3.2.

Table 3.2. Threshold Levels for Declining Revenue Risk Measure

Rating	Change in Revenue Moving Average from Two Years Ago
G	$R_c(t) < 1\,\sigma$
Y	$2\,\sigma < R_c(t) \leq 1\,\sigma$
O	$3\,\sigma < R_c(t) \leq 2\,\sigma$
R	$R_c(t) \leq 3\sigma$

Improvement options. The initial risk-rating thresholds were set by our team's expert in corporate financials. Once a larger data set is available, we may want to reset these thresholds based on statistical analysis of the period in question and our assessment of the stability of the DIB at the time. For example, if the DIB is relatively healthy at the time, then fixing the thresholds based on the standard deviation approach discussed earlier would yield a relative ranking, but its fixed nature would give us an ability to know when the industrial base is entering a worse (or better) period.

Jump in Federal Workload

Motivation. A significant increase in workload may indicate a capacity risk for a contractor (especially in the short run) and possibly a cost growth risk if they need to make new capital investments to increase capacity. Although data on corporatewide bookings of new orders and contracts are hard to obtain, we do have readily available data on federal contract obligations through FPDS-NG. Thus, such federal workload will give a partial insight into jumps in workload.

Approach. We used a three-year moving average to smooth the data and reduce false positives of jumps in obligations from companies winning large contracts. The DoD Green Book (Office of the Under Secretary of Defense [Comptroller], 2018) indicates that, on average, most DoD obligations are expended (paid) within the first two to three years, so we used a three-year moving average to smooth out the execution of large contract awards.

1. Extract total annual federal obligations $O(t)$ from FPDS-NG.

2. Calculate the three-year moving average (O_{ma}) of total annual obligations (with a floor of $1 million to avoid noise for minimal company activity).

$$O_{ma}(t) = max(\$1M, (O(t-2) + O(t-1) + O(t)) / 3)$$

3. Calculate the percentage change O_c in moving average from one year ago.

$$O_c(t) = (O_{ma}(t) - O_{ma}(t-1)) / O_{ma}(t-1)$$

Rate the *jump in obligations* risk using the fixed thresholds shown in Table 3.3. These fixed thresholds (10 percent, 20 percent, and 50 percent) are close to the one, two, and three standard deviations in the current data sample.

Table 3.3. Threshold Levels for Jump in Obligations Risk Measure

Rating	$O_c(t)$
G	$O_c < 10\%$
Y	$10\% \leq O_c < 20\%$
O	$20\% \leq O_c < 50\%$
R	$O_c \geq 50\%$

Improvement options. Another potential approach is to extract the backlog of work from a company's SEC 10K filings. This is only possible for public companies in the United States and would involve somewhat more-sophisticated text analysis to identify and extract backlogs in free text sections of the filings.

As for the threshold levels, we could update them once additional data are available in the system to ensure that they align to the population's one, two, and three standard deviations. We could also set them to be dynamic and adjust with the population over time, or we could fix them with the current population distribution. The latter has the advantage of allowing for the number of risks triggered to increase as the revenues decline across the defense industrial base.

Financial Category

Financial Metrics

Motivation. Various health metrics are in common use by the financial markets to monitor and assess a company's financial health (e.g., to assess financial risks and inform stock purchasing). Although the DoD is not involved in investment decisions, these data should be useful to assess whether a company is in a weakened financial condition and thus may warrant attention from an industrial base or contractor performance perspective.

Approach. For the prototype, we obtained access to S&P Global's CapIQ data (S&P Global, 2017)[7] through a DAF license. As with other vendors, CapIQ provides structured financial health metrics extracted from various sources (e.g., from SEC filings for public companies in the United States). Our initial approach is to examine the Credit Health Panel for a working sample of companies. That panel reports 24 specific metrics in three areas: operational, solvency, and liquidity.[8] We then compared a specific company's ratings with those of the peers identified by CapIQ. For example, CapIQ identified 21 domestic and international peers of the Boeing Company. We applied the deviation approach discussed earlier to determine whether the company was at least one standard deviation worse than its peers to score its relative risk. To obtain an overall relative financial risk measure, we employed the approach outlined later in this chapter to combine risk scores.

Sources considered for our financial analysis criteria were primarily S&P Global's CapIQ (S&P Global, 2017), but we also evaluated Fitch Ratings (undated); Lermack (2003); Moody's Investor Service (2014); U.S. Government's Pre-Award Survey of Prospective Contractor Financial Capability, Standard Form 1407 (U.S. General Services Administration, 2014); Standard & Poor's (2014); the SEC's Office of Credit Ratings Updated Investor Bulletin on The ABCs of Credit Ratings (SEC, 2017); and various financial websites.

Our analysis of these results indicated that it would be useful to focus on the more important of these 24 financial metrics, so we selected five to highlight and discounted or disregarded the others. We selected our primary financial measures of operational, solvency, and liquidity based on our financial software license from S&P Capital IQ. The selected metrics are traditional financial measures that evaluate a company's performance or rate of return and profit margin (operational); its ability to meet their debt (net debt/EBITDA) and interest expense obligations (EBITDA/interest expense that measure solvency); and its available capital on hand (liquidity). The financial measures selected were based on information that companies included in their quarterly and annual SEC filings, company presentations to investor analysts, and experience of our subject-matter experts on the team. However, these 24 metrics can be weighted to match the user's priorities or preferences.

---

[7] Note that there are other related business intelligence products on the market. We are using CapIQ as a proof of concept, not as an advocacy of a particular product.

[8] The 24 financial metrics are as follows:

- *Operational*: Total Revenue, Total Equity, Return on Capital (%), Recurring Earnings/Total Assets (%), Net Working Capital/Revenue, Asset Turnover, Intangible Assets/Revenue, Net Working Capital/Total Assets, Payables/Receivables, Management Rate of Return (%), Gross Margin (%), and Earnings Before Interest, Taxes, Depreciation, and Amortization (EBITDA) Margin (%)
- *Solvency*: Funds from operations (FFO) Interest Coverage, EBITDA/Interest Expense, FFO to Total Debt, Net Debt/EBITDA, Total Debt to Capital (%), Total Debt/Total Liabilities (%), and Total Debt/Revenue
- *Liquidity*: (FFO + Cash) to Short Term Debt, FFO to Gross Profit, Basic Defense Interval (days), Current Ratio, and Quick Ratio.

Improvement options. It would be useful to allow the user to change the weights for the different metrics through the prototype's user interface. We may also want to develop a PSC-based or NAICS-based algorithm[9] to develop the list of peers for companies so we can expand it beyond those that CapIQ has identified.

Customer Base: Declining Revenue

Motivation. If a company is experiencing a declining trend in total revenue, then there is potential that the company's customer base is declining and may be a risk.

Approach. This is the same relative risk measure included in the "Cost Category" section earlier in this chapter.

Improvement options. See ideas for declining review in the earlier "Cost Category" section.

Stability Category

Excluded Contractors

Motivation. The federal government maintains a centralized database called SAM.gov that lists companies approved to do business with the federal government. Some contractors are precluded from having government contracts, however, for a variety of reasons: being an individual barred from entering the United States, being an entity that has violated national security protocols, and being entities that have been convicted of tax fraud (see Federal Contractor Registry, 2019). Normally, a contracting officer should check SAM.gov to see whether a potential contractor is approved, but we include it here so that it is available with other contractor risks.

Approach. Here we simply looked up each contractor in question in the SAM.gov data to determine whether it is "excluded."

Improvement options. None identified. Exclusion is a fairly clear problem with a contractor.

Recent or Pending Mergers and Acquisitions

Motivation. M&As can be disruptive to a company's operations, finances, workforce, and performance.[10] Some M&As, of course, can be very beneficial by bringing new capabilities. Others can trigger a period of increased problem solving and internal distractions. Thus, a first-order measure is to indicate the existence of a recent or pending M&A so the PM knows about it.

---

[9] NAICS is a standard classification system used by the United States, Canada, and Mexico (see Office of Management and Budget, 2017).

[10] We used several sources for the types of problems that may be encountered post acquisition, including Deloitte, 2009; Deeb, 2016; Seth, 2019; and Merger Integration, 2019. In addition, we used the industry M&A experience of our subject-matter experts on the team.

Potential data sources include news feeds (e.g., LexisNexis, news wires, online news engines) or news feeds from financial information providers, such as CapIQ, FactSet, or Eikon. CapIQ, for example, has a structured M&A data service (S&P Global, 2017).

Approach. The prototype consumes a data feed from CapIQ identifying all M&A activity that has taken place within the last two years (a total of 2,063 mergers). The application then identifies which of the companies involved, either as a buyer, a seller, or as the organization being acquired, are Air Force contractors and marks them as having a slightly elevated risk level (a risk level of yellow).

Improvement options. This risk could incorporate further improvements which might predict the overall difficulty of successfully prosecuting the merger. For example, the relative size of the companies might have an effect on the specific nature of the risks. A merger of relative equals or an acquisition of a substantially different line of business could be more distracting for management than an acquisition of a small company well aligned with the company's previous core competencies. Alternatively, when a small company is bought by a larger company, there may be a larger risk that its existing commitments go unnoticed by the management of the combined conglomerate. Finally, integrating news sources to judge the overall risk of the acquisition may result in further improvements. A merger, even of a relatively small company, that is deemed essential to the combined company's future is very likely to distract upper management. Similarly, news sources may provide early warning of implementation difficulties in successfully combining the companies.

Lawsuits Pending

Motivation. As with M&As, pending lawsuits may disrupt the internal operation of a company, distract its workforce, and introduce financial or even existential risks. Of course, the details of the lawsuit should greatly influence the risks. Potential data sources include news feeds (e.g., LexisNexis, news wires, online news engines) or news feeds from financial information providers, such as CapIQ, FactSet, or Eikon. CapIQ, for example, has a filtered news feed that we are using in the prototype for the cross-cutting news risks discussed later in this chapter (S&P Global, 2017).

Approach. This idea was not implemented in the prototype, but the basic approach is to alert that a pending lawsuit exists to avoid ignorance and trigger any due diligence by the PM. A basic approach is to use a simple keyword filter to identify lawsuits in the data sources used. Because the existence of a lawsuit may or may not indicate significant issues, the initial implementation could be to just alert it as a yellow risk.

Improvement options. This risk measure is a candidate for an expanded prototype. Beyond simply alerting to the existence of a pending lawsuit, more sophisticated legal analysis might be considered, although such analysis probably involves significant R&D by itself.

Supply Chain Category

Corporatewide Supplier Risks

Motivation. Many (if not most) prime contractors have suppliers or subcontractors of various sorts. They depend on those lower-tier suppliers to perform on their prime contracts. Therefore, risks in a company's supply base can introduce performance risks for the company itself.

Approach. Here we use available data to identify relative contractor risks of lower-tier suppliers in the same ways we do for the prime contractor. Thus, all the relative risk measures discussed in this report apply recursively to the lower-tier suppliers. We do this by identifying potential relative risks for all companies for which we have data, then identify prime-subcontractor relationships to recursively record supplier risks for primes. We found that CapIQ identifies corporatewide suppliers and therefore can be used to link suppliers to primes. A total of 4,866 suppliers have been identified this way. As dependencies identified in this way may or may not be relevant to a particular Air Force program or contract, this approach does highlight an area of potential risk that requires acquisition professional due diligence.

Improvement options. As with many of the uses of CapIQ or other financial information providers, implementing such risk assessments conceptualized here require computer-to-computer information sharing through application programming interfaces (APIs). Thus, beyond manual proof-of-concept examinations, we are seeking to obtain an API to further test the utility of this approach. Also, other information providers (besides CapIQ) collect and structure this kind of corporatewide supplier relationship data. A further examination of information source options could be beneficial if the prototype is developed into a fully operational system for the DAF or DoD-wide.

Contract-Specific Suppliers Department of the Risks

Motivation. As with corporatewide suppliers, any lower-tier risks may flow to prime contractors and affect their performance on DAF contracts. Ideally, we would want to know whether those suppliers are directly involved in a specific contract to help identify whether the risk is truly relevant to the DAF and whether it effects larger constructs, such as acquisition programs.

Approach. The recursive risk approach is similar to that for corporatewide supplier risks, except that the prime-supplier relationship is based on different data. One data source is the subcontractor relationships reported to FSRS and available publicly through USAspending.gov. A total of 708 contract-specific subcontractors have been identified. Subcontracts worth more than $25,000 are required to be reported to FSRS.

Improvement options. Anecdotes indicate that FSRS data may not be complete. If so, and if FSRS data become critical for automated data analysis of the industrial base as discussed in this report, then the DoD may want to explore ways to improve data reporting and quality in FSRS.

Influence Category

Insignificant Customer: Department of the Air Force or Department of Defense (Influence)

Motivation. If the DAF or the DoD is only a small portion of a contractor's business, then there is a risk that the Air Force and the DoD will have difficulty influencing the contractor to perform well, address problems, and focus their best minds on Air Force or DoD contract work.

Approach. Here, we compare the running three-year moving average of obligations from FPDS-NG against the three-year average of the company's total revenue to estimate the percentage of work as an Air Force prime contractor or a DoD prime contractor. This will only show work as a prime contractor and will not include any work as a subcontractor, but it should provide some perspective on influence. The algorithm equations for three-year moving averages of obligations is similar to that for the three-year moving averages of revenue discussed earlier for the risk measure for a Jump in Federal Workload. Our initial implementation used the thresholds shown in Table 3.4.

Table 3.4. Threshold Levels for Insignificant Customer

Rating	Percentage of Revenue
G	$R_p \geq 10\%$
Y	$R_p < 10\%$
O	N/A
R	N/A

Improvement options. Further analysis of the population is needed to set the thresholds at levels that reflect reasonable levels while avoiding excessive alerts. There is no clear theoretical level at which attention is reduced, so this might be change to a user-adjustable level. Also, the data could be expanded to federalwide data from FPDS-NG to determine whether the federal government has much influence on a company. We could also expand the calculation to include subcontractor data available from FSRS (via USAspending.gov). Finally, influence may change depending on factors within subsets of a company (e.g., a large portion of a small division in large company may still retain influence).

Performance Category

Past or Current Contractor Performance

Motivation. A contractor's performance on past or current efforts should be a useful risk indicator for future performance. Past performance, for example, is often used in source selections as a ranking indicator. Also, there may be ways to associate current performance issues on other contracts for similar work. For example, if Contractor X is having trouble performing on three Army software development efforts, then there is some reason to be concerned about Contract X's software development efforts for the Air Force.

Approach. Our current approach is to extract contractor performance on Major Defense Acquisition Programs (MDAPs) from DCMA PARs. These quarterly reports provide aggregated program-level green, yellow, or red ratings for DCMA's contract performance assessments (PAs), and management assessments (MAs). The prototype extracts these program-level ratings and displays them on the individual program pages, where such data are available.

Improvement options. In addition to these semistructured PARs, DCMA has deeper data on various contractor aspects in their eTools databases. Exploring and including portions of those databases may be very valuable.

Also, the federal government annually records contractor performance in CPARS on all larger contracts (thresholds are specified in Federal Acquisition Regulation [FAR] 42.1502). Thus, if access can be obtained (and data protected appropriately), CPARS data could serve as a significant indicator of potential contractor risk based on past and current performance.

Experience in Key Work Area

Motivation. This is the same relative risk measure included under the Workforce category earlier in this chapter. For the workforce, it may indicate whether a company's workforce has experience in the PSC subset or whether the size of the workforce in the PSC subset may be insufficient because of a large increase in work in the PSC subset. It may also reveal potential performance risks on the contract because of the newness of the work area or a large growth in the work area.

Approach. (See the algorithm under the Workforce category.)

Improvement options. (See the discussion under the Workforce category.)

Prior Experience Working for the Air Force or Department of Defense

Motivation. Having no prior contracts for the Air Force or the DoD introduces a learning curve. This risk measure indicates whether there is a record of a past contract with the Air Force in the seven years of DoD data downloaded from FPDS-NG.

Approach. Here we give the contractor an orange rating if the contractor does not appear in the recent data (since FY 2013) and green otherwise.

Improvement options. It would be useful to expand the search to include all federal work, because knowing that the contractor has not dealt with federal contracting at all gives a deeper insight into potential risks. Also, it would be useful to assess whether all prior contracts (if any) were fixed price and did not require cost accounting reports to the DoD (e.g., firm fixed-price) or if at least some involved cost accounting reports to the DoD (e.g., cost-reimbursement contracts and incentive contracts). Lack of experience in the latter is a potential risk worth noting.

A different issue that could be addressed further is to expand on the U.S. Air Force's DUNS history data to help correlate and track companies through M&As and joint partnerships and ventures. Because the Air Force contractor base frequently engages in M&A activity, considering the prior work experience of acquired companies and subunits might eliminate some false-positive risks when a company has acquired a unit with prior experience in a particular line of business.

Security

We did not implement any security risk measures because of a lack of accessible and usable data sources. Table A.10 in Appendix A outlines some potential risk measures and conceptual data sources, but more investigation is needed. Consider cybersecurity. The most promising approach may be to leverage ongoing efforts to identify cybersecurity plans or maturity once those efforts are implemented. For example, the Defense Acquisition Regulation Supplement clause 252.204-7012 requires contractors to document their implementation of National Institute of Standards and Technology Special Publication 800-171 in a System Security Plan (DoD, 2019). An archive of such plans would provide an indicator of potential cybersecurity risks. Alternatively, efforts are underway to develop a Cybersecurity Maturity Model Certification (CMMC) (Office of the Under Secretary of Defense for Acquisition and Sustainment, undated; Barnett, 2019). Again, if data were made available on CMMCs, then companies (including lower-tier subcontractors) that lack a CMMC could be deemed to have higher cybersecurity risks. Unfortunately, these regulations and certifications are still being implemented and developed, and no central data archives yet exist, illustrating how security measures are proving more difficult than other risk measures.

Infrastructure

We did not implement any infrastructure risk measures because of project time constraints. However, Appendix A outlines some potential risk measures and conceptual data sources. The most promising approach identified is to assess the stability of production quantity.

Production Stability

Motivation. One idea for infrastructure risk (outlined in Table A.11 in Appendix A) is to look across multiple production lots on a program for instabilities in quantity over time. Ideally, quantity would be relatively stable so the contractor can optimize the production line and workforce. Otherwise, additional costs may be incurred if the contractor had to ramp up or down capabilities.

Approach. This idea was not implemented in the prototype, but the basic approach is outlined earlier.

Improvement options. This risk measure is a candidate for an expanded prototype because data on production quantity by program and contracts are readily available in PMRT and DAVE.

Capacity

Jump in Federal Workload

(This is the same relative risk measure included in the "Cost Category" section earlier in this chapter.)

Production Capacity (Economic Order Quantity)

Motivation. Another idea for capacity risk is outlined in Table A.12 in Appendix A. EOQ in the DoD is the level of quantity assessed to be optimum for the current production capacity. Although EOQ is an approximation given that multiple production lines are often involved in producing a system, EOQ provides an assessment of optimality that could be reviewed against actual order quantities to assess risk.

Approach. This idea was not implemented in the prototype, but the basic approach is outlined above.

Improvement options. This risk measure is a candidate for an expanded prototype because the calculations are straightforward and both quantity and EOQ are often available for MDAPs in PMRT and DAVE/DAMIR.

Future Category

No risk measures that relate to the future of each contractor were implemented in the current version of the prototype, but we outline ideas in Table A.13 in Appendix A and briefly describe one measure below that has more-accessible data.

Low Government-Funded Research and Development

Motivation. One type of R&D funding for which data are readily available is government-funded R&D. If a company is earning low levels of R&D funding relative to their peers in their industry, then there may be an increased risk that they will not be as viable as their peers in the long run—especially if their peers tend to survive on government investments (as is usual for major weapon systems in the DoD). Trend data may also be useful. For example, flagging companies that have historically earned R&D funding in specific areas (compared with their own earnings) have been losing ground; this might indicate a reduction in competitiveness, a workforce issue (R&D staff preservation or skill mix) if the reduction is due to a shift from R&D toward production, or a general degradation of building future areas of expertise. Thus, a relative measure of R&D funding levels should produce a useful relative risk indicator for subsequent due diligence.

Approach. This idea was not implemented in the prototype, but the basic approach is to extract R&D funding obligations (either in total or by PSC area) and compare with their peers (e.g., those that operate in the same PSC area). Statistical measures, such as a standard deviation, might be employed to identify those with lower R&D funding than the others in their population. Trend data may also be useful to compare directly with a company's past to see whether it is losing ground.

Improvement options. This risk measure is a candidate for an expanded prototype. Other potential data sources might be considered, such as proprietary data on independent R&D (IR&D) investments and progress as outlined in DTIC.

Cross-Cutting Risk Measures

News Alerts: Companies and Programs

The News Alert feature seeks to provide users with information about major events affecting contractors working on their programs. The information provided is in the context of other risk conditions that have been identified by the application. It builds on CapIQ's existing per-company news feature. Additional news feeds, such as ProQuest or Factiva, could also be purchased to add additional newspapers, journals, or industry-specific magazines.

This feed of news stories is then filtered to look for acquisition-relevant keywords to increase the relevance of the feed. Only stories that contain at least one of the keywords listed in Table 3.5 are shown in the prototype. This keyword list was developed using the project team's expertise in acquisition. When displayed, these keywords are highlighted to help the reader see what risk may be indicated in the story.

Table 3.5. Keywords for News Story Filter

Attrition	Fail*	Loss	Requalif*
Audit*	Foreign component	Lost	Resign*
Bad	Foreign supplier	M&A	Restructure
Bankruptcy	Fired	Merger	Rework
Breach	Firing*	Penalty	Scandal
Charge	Hack*	Poor Performance	Schedule growth
Close*	Harassment	Price growth	Schedule slip
Compromis*	Halt	Price increases	Scrap
Cost growth	Hearing	Problem	Scrap rate
Cost increases	Hostile	Outsource*	Security breach
Counterfeit	High* overhead	Overhead increases	Security problem
Court	Immatur*	Overrun	Shortcoming
Damage	Incident	Phishing	Shortfall
Death	Inexperience*	Poor	Slip
Default on	Inflate	Privacy	Stop work
Deficien*	Inflated	Protest	Suit
Deficit	Injury	Rebuild	Surplus
Delay	Insider threat	Reconstruct	Takeover
Denied	Instability	Redesign	Tampering
Denial	Insufficient	Reorganiz*	Theft
Destroy	Judge*	Repair	Threat*
Destroyed	Lack	Replace*	Turnover
Destruction	Late	Replan	Unstable
Disbar*	Lawsuit	Reputation	Vulnerab*
Disqualified			

NOTE: Asterisks are wildcards representing zero or more characters.

Combining Relative Risk Measures

To combine relative risk measures, we first converted the ratings to a numerical score and then used various functions to combine those scores to obtain an overall relative risk measure.

Scoring Relative Population Risks

For our initial prototype, we used a simple algorithm to convert a single relative risk value to a number. We used the associated population fraction should the population distribution be Gaussian:[11]

$$Relative\_score = 1 \: / \: Population\_fraction$$

---

[11] We realize that we do not know whether the distribution is Gaussian, but this simple approach gives what we need: a strongly nonlinear scoring that heavily emphasizes companies that are three standard deviations away from the mean over those that are two standard deviations away (and similarly compared to those that are one standard deviation away). Further research and effort could be made in subsequent iterations of the prototype to empirically assess the actual distributions and associated fractions, but this simple approach was useful for a first iteration to get an operational prototype functioning and allow us to begin assessing these finer points.

Here the score is simply the inverse of the fraction, as illustrated in Figure 3.1. For example, if the score was yellow, then that score would be associated with standard deviations between +1 and +2. In a Gaussian distribution, about 13.6 percent of the population would be in that range. Thus, the score is *1/.136 = 7.4*. Of course, we could use the exact standard deviation value directly rather than the grouping value to obtain a more precise score, but this illustrates the concept. In addition to reflecting how far the risk measure is from the population, this yields a nonlinear scoring value that reflects extreme outliers, which is our goal in this step.

Figure 3.1. Relative Risk Rating and Score

Relative score = 1 / Fraction

		Fraction	Score
G	$x < 1\sigma$	84.2%	1.2
Y	$1\sigma \leq x < 2\sigma$	13.6%	7.4
O	$2\sigma \leq x < 3\sigma$	2.1%	47.6
R	$3\sigma \leq x$	0.1%	1000

Combining Scores Within a Category

Now that we have converted an individual relative risk measure to a numerical score, we can use simple functions to combine multiple measures with a category and across categories to obtain summary risk scores.

Within a risk category, the preponderance of multiple risk measures should reflect the strength of the potential concerns and the confidence that there may be a risk that is worth due diligence by the user. In other words, if multiple measures in a same risk category indicate that there may be a risk, then the confidence should be higher. Thus, we took the average (arithmetic mean) of the available risk measures within the category. In some cases, we only have one measure, so the category score is simply that single measure score. In other cases, we have many measures, so the average helps us understand how strong the risk is when we look at it in multiple ways within a category. The right side of Figure 3.2 illustrates this concept for the large number of financial health measures available. In this case, we have 24 measures across operational, solvency, and liquidity. All are green except one orange and one yellow rating. Here, the average is yellow because we have a nonlinear value function for combining ratings. We do not lose the fact that there are two measures above green, but the preponderance of measures indicates that these are localized to just two of the 24 measures and thus lower than, for instance, the highest relative risk (orange).

Figure 3.2. Combining Relative Risk Measures Within and Across Categories

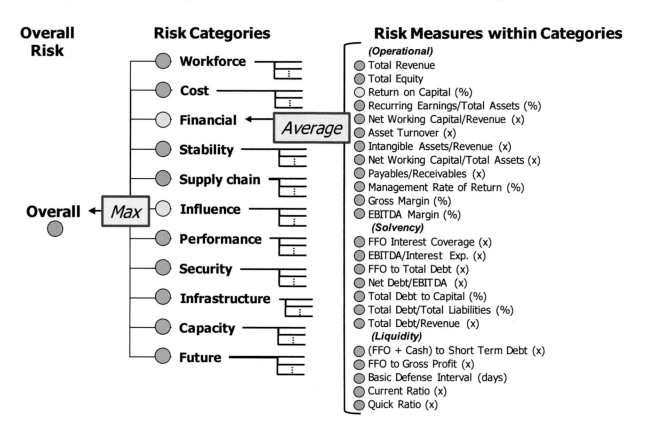

Once we have a combined score within a category, we simply applied the same breakpoints in the original scoring scheme to assign a color to the combined score (see Figure 3.3). Anything above 1.2 is green. Anything above that and up to the nominal score of 7.4 is yellow, and so on.

Figure 3.3. Relative Risk Rating Based on Combined Scores for Multiple Measures

Combined score	Rating
1.2	G
1.3–7.4	Y
7.5–47.6	O
≥47.7	R

Other functions can be used to combine scores within a category, depending on how many risks are triggered and how many the user wants revealed. For example, one might prefer to use a *maximum* function so that any risk (the worse risk) is propagated up from the category level. If the maximum were taken among the relative risk measures in Figure 3.2, this would mean that the Financial category value would be orange instead of yellow. Therefore, as with the scoring functions, the choice of a combination function has a significant effect on the revelation of potential risks and thus the user's perception of potential issues. Further research and user input are needed to understand what the best function might be or whether a set of selectable functions should be made available through the user interface. Ideally, we would understand both the relative importance of different measures within a category and how those measures interrelate, but that level of understanding is beyond the scope of this initial stage of exploratory work.

In addition to taking the average of the relative risk scores, we can weigh the relative importance of the available measures. In the case of financial health measures, we implemented weights to prioritize a few that our financial expert on the team deemed the most important (see earlier discussion). Such weightings could conceptually be set by the user based on what risks she or he is most concerned about.

Combining Scores Across Categories

To combine relative risks across categories, we used the maximum function because we want the user to see the worse potential risks across the categories. This initial choice was driven in part because the data we used initially had few relative risks above green, and we did not want the risks in a single category to be washed out by other categories. Of course, other functions could be easily implemented (e.g., using the average, as we did within each category, or perhaps a simple summation function).

The left side of Figure 3.2 illustrates how the *maximum* function reveals the worse risk. In this notional example, the Supply Chain is a relative risk of orange, so the overall risk is shown as orange.

Although we use colors to rate the risks and make it easy for the user to quickly see higher risks, the system retains and displays the underlying scores. Thus, a combination *yellow* score that is supported by two yellows has a higher risk than one that is supported by a single yellow.

Observations

In summary, the risk measures discussed in this chapter combine available data to indicate when a company may have an increased risk when compared with the values for other companies. Although the approach usually cannot definitively identify that a risk exists, those measures are objectively applied to all companies and constitute a way to reduce the number of issues for further due diligence. The next chapter describes the prototype implementation of the approach, but further work is needed to evaluate the sensitivity and utility of the proposed measures once the prototype is matured and more data are added.

4. Prototype Architecture

Intended Uses

The software prototype designed and implemented for this project was intended to serve two primary categories of users and their business needs. The first user category consists of DAF PMs or portfolio managers who oversee one or more Air Force programs. The second user category consists of Air Force analysts or other personnel assigned to the headquarters component who must keep track of the overall state of all Air Force acquisition programs.

Given these two categories of users, the application has been primarily designed around two common uses for these personnel. In the first, an analyst (for example, in the Office of the Deputy Assistant Secretary for Acquisition Integration [SAF/AQX], Office of the Assistant Secretary of the Air Force for Acquisition, Technology, and Logistics) wishes to quickly identify the most problematic Air Force contractors compared with peers and to discover whether there are any new problems for closer examination. After logging in, this analyst will be brought to the All Contractors page. This page lists contractors whose relative risk level has increased since the previous evaluation period, followed by a list of contractors whose relative risk level has decreased since the previous evaluation period. This allows the analyst to quickly discover contractors with new risks or points of concern that might require further investigation to understand. After this, the application lists all of the known contractors, from the riskiest to the least risky, allowing an analyst to quickly determine which Air Force contractors present the greatest relative risk at the current time. The analyst can click on any of the listed contractors to see more information about the nature of the risks associated with that contractor. The analyst can also search for a particular contractor of interest by typing a partial match for its name into a search box.

In the second usage, an acquisition professional with responsibility for Air Force programs wishes to understand the state of the overall acquisition portfolio. Two types of pages facilitate this usage. The Single Program page will list all of the information this prototype knows about an individual program. In particular, this includes available information on any contractors that work on the program, any subcontractors that are known to work on the program, the relative risk of those contractors, and any recent PAR evaluations for the program. The page will also list recent news stories about the program that are relevant to acquisition personnel. If one of the contractors on their program has an elevated level of risk, the user can see details about the identified risks for that contractor by clicking on their name and reaching the associated Single Contractor page. This page will list the overall risk for the contractor in each category of risk and explain the specific risk conditions that have been detected for this contractor. It will also list all known subcontractors for this contractor (regardless of which program those subcontractors work on) and any recent news stories about this contractor that are relevant to acquisition personnel. Both the Single Program page and the Single Contractor page are reachable from the search box in the application or by clicking on any reference to the contractor or program from the All Contractors page or the All Programs page.

Architecture Overview

To support these uses, the relative contractor risk prototype consumes data from a variety of original sources, including restricted government sources (such as PMRT), open government sources (such as FPDS-NG, the BLS website, and USAspending.gov) and open or for-fee commercial sources (such as job posting websites and aggregators of financial data). Each of these data sources feeds a data pipeline in the prototype architecture. Pipelines have two primary components: data parsers and risk calculators.

As a first step in these pipelines, parsers specific to each data source clean and normalize the data into standard formats. This serves several purposes. To begin, it removes data that may be missing critical fields or fail other validity checks. It also allows for data that originate from different data sources to be associated with shared concepts or data constructs. For example, correctly identifying that data from different data sources to describe the same company or Air Force program can be a challenge when the input data lack a unique identifier to confirm the match; parsers must look for near-matches between the input data and known companies or programs to determine whether the input data represent something new or can be matched to a known entity. Finally, this also allows the application to add, remove, or change data sources that provide similar types of information without the need to alter or rewrite any part of the application beyond the data parser specific to the data source in question.

Once the data has been parsed and cleaned, a different set of background software examines the data to detect any risk conditions for Air Force contractors or subcontractors. These risks can use data from one or more data sources, depending on the specific nature of the risk. Data pipelines are not run continuously; they only need to be run when new data are received. This can vary based on the nature of the data. For example, financial data are typically updated quarterly when SEC requirements force corporations to update their investors and potential investors about the state of their business. Other data, such as open job postings, could be updated on a monthly or even daily basis if desired. Once complete, the risks generated by these pipelines are stored in a PostgreSQL database, which other portions of the application can reference at any time.

As a final step, once all of the risk modules have completed and any newly identified risks have been stored, a final software task calculates the relative risks of all known Air Force contractors within each of the 11 risk categories and an overall relative risk rating for each contracting company. The algorithms to combine relative risk measures are described in Chapter 3.

Each of these data pipelines and risk calculators operate independently from the final significant module within the relative contractor risk prototype: the user interface. Users can interact with the application through a standard web interface. For the prototype, the user interface has been built using the model-view-controller pattern and can be deployed on a standard Microsoft Internet Information Services web server. Figure 4.1 shows the end-to-end architecture of the application.

Figure 4.1. The Prototype's Architecture

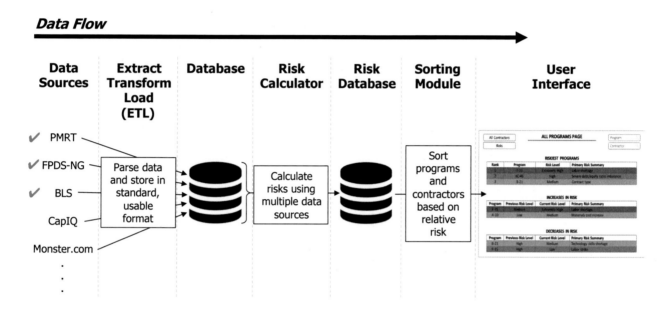

User Interface

Users of the relative contractor risk prototype interact with the data and results through a web interface. As discussed earlier, it focuses on four primary views: an All Contractors page, numerous Single Contractor pages, an All Programs page, and numerous Single Program pages. Information on each of these pages is often linked: For example, the Single Contractor page displays information both about the specific contractor of focus for the page, other contractors with whom this contractor has some kind of relationship, and Air Force programs that the contractor works on. Users can easily navigate between these pages, as shown in Figure 4.2.

43

Figure 4.2. User Interface Structure for the Prototype: Contractor and Program Views

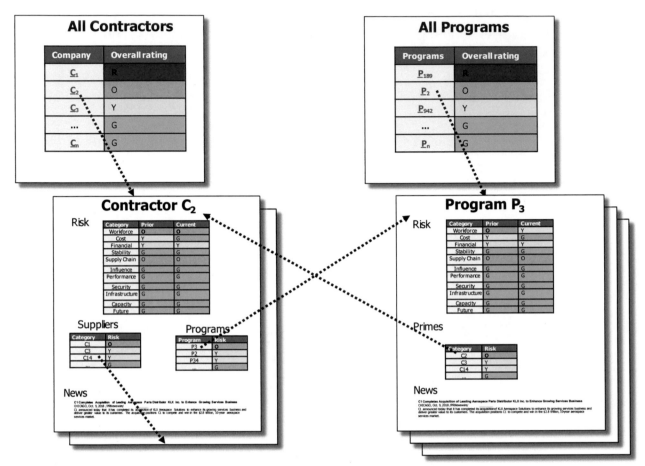

NOTES: The arrows illustrate linkages in the prototype between views. Thus, the Program P_3 element on the Contractor C_2 page points to the Program P_3 page. The arrow in the lower left from Supplier C_{14} points to the Contractor C_{14} page lower in the stack.

Additionally, every page in the web application has some global controls for navigating the prototype. Each page has weblinks to the All Contractors page and the All Programs page that allow the user to quickly return to an overview of all contractors or programs from a more specific search. Additionally, every page has a search box that can be used to navigate to the webpage for a specific contractor or specific program. The box will display suggestions for companies and programs as the user types based on the text provided. Once the user sees a topic of interest, the user should select the appropriate option in the list and wait for the program to navigate to the appropriate webpage.

All Contractors Page

The All Contractors page is designed to quickly give the user an overview of which Air Force contractors have relevant information that an acquisition professional might want to investigate. Figure 4.3 shows a snapshot of the page with notional data. At the top, the "Relative Risk Increases" table displays contractors whose overall risk level has increased since the past evaluation period for contractor risks. Each of these contractors have had some new risk condition identified, which increases their overall level of risk from its previously determined level; consequently, acquisition personnel may be interested in understanding exactly what has changed and what could potentially rectify the situation. A user can drill down on each cell in the table; clicking on one will take the user to the Single Contractor page for the company in question. Similarly, the "Relative Risk Decreases" table shows contractors whose overall risk level improved since the past evaluation period. The user can also drill down on each row in this table to investigate a particular contractor more deeply.

Next, the "Risk Level Breakdown" table shows how many government contractors have been classified at each risk level. Green is the lowest relative risk category, and red is the highest relative risk category.

Finally, the "Current Contractor Relative Risks" table lists every contractor for which the application can evaluate at least one risk condition—from the riskiest to the least risky. A Relative Risk Score is displayed to give the user an idea of how large a gap exists between two particular contractors. This score is not intended to focus the user on any particular score or number; rather, the scores are displayed simply to give the user a general understanding of the degree of gap between one risk level and another risk level, especially if the scores are close but the risks have different color assignments. This should help a user triage how many risks and contractors they have time to inspect and allow them to focus on an appropriate number of contractors and programs.

Figure 4.3. Prototype User Interface: All Contractors Page

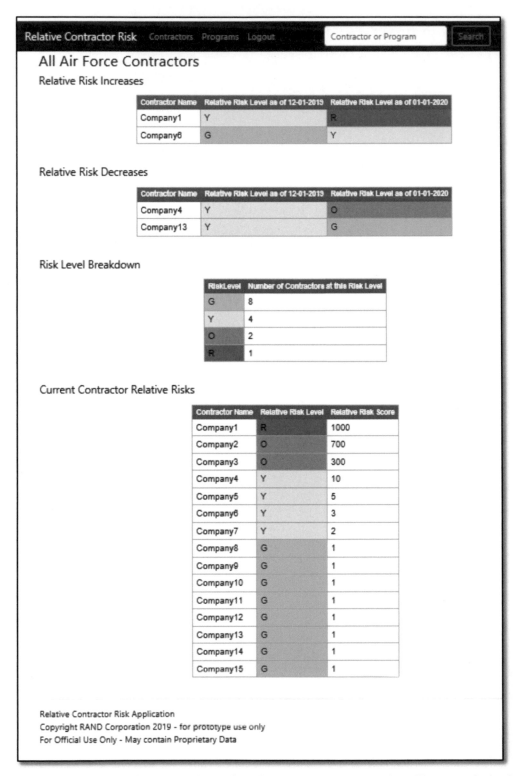

NOTE: This notional screenshot is from an early version of the prototype with test data. The numerical values in this screen are not illustrative of operation.

Single Contractor Page

The Single Contractor page is designed to provide the user with an overview of a particular government contractor and the known risks that have a possibility of affecting their performance. Figure 4.4 shows a snapshot of the page with notional data. At the top of the page, the current "Contractor Relative Risk Level" is identified and highlighted in the appropriate color. This score represents the amalgamation of all risk factors for this contractor considered relative to other government contractors. It is calculated by the Rank Contractors job. For context, a coding key is provided to explain the four relative risk levels: red, orange, yellow, and green.

Figure 4.4. Prototype User Interface: Single Contractor Page

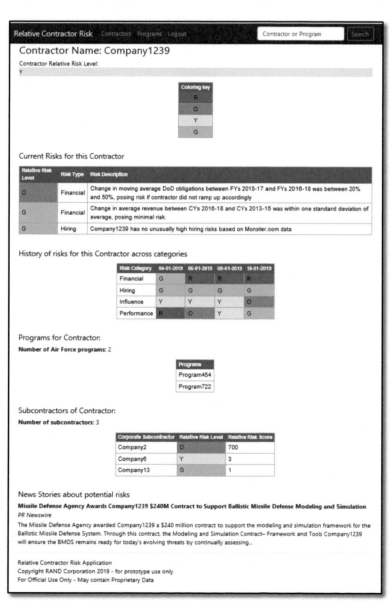

Below the overall risk level, the "Current Risks for This Contractor" table provides specific information about every individual risk that has been identified for this contractor. Each risk factor is assigned to one of the 11 risk categories identified by this project. Each risk factor can be present to a greater or lesser degree, with the "Relative Risk Level" column visually identifying the degree of risk along the red/orange/yellow/green scale. The "Risk Description" column provides a human-readable explanation of what this specific risk is measuring and why this contractor is or is not at an elevated risk level. This will allow Air Force personnel to follow up appropriately. In some cases, Air Force personnel may be able to help mitigate the risk conditions by working with the contractor in question. In others, a deeper look at the specific characteristics of the contractor and program may find additional factors that heighten or ameliorate the risk condition. In all cases, this table is intended to give Air Force personnel insight into risk conditions that they may not have time to actively monitor or allow them to identify concerns about not having access to data sources that would alert them to the risk condition.

Below this, the application displays a table showing how the level of risk has changed for this contractor in each category of risks over time. This will provide useful context to help the user understand which kinds of risks have been increasing over time, which seem to be improving, and which have stayed relatively constant.

Next, the application displays a table showing all the Air Force programs that this contractor works on. It also displays the total count of how many programs this company is a contractor or subcontractor on. The user can drill down on each row of this table to investigate further; clicking on any cell of the table will take the user to the appropriate Single Program page to review more information.

The last table displays all the companies that are known to be subcontractors of this company. The application may or may not know which specific program the two companies collaborate on; consequently, all known subcontractors are displayed on this page, while subcontractor relationships that are specific to a particular Air Force program are shown on the Single Program page. The user can drill down on this table by clicking on any individual row. This will take the user to the Single Contractor page for the subcontractor they selected. This table also displays the overall contractor risk level accessed for each subcontractor.

Finally, the application will display news stories of interest about this particular contractor. These news stories originate from a variety of publications and are filtered to only include stories of relevance to this company and stories that contain one or more words from a list of acquisition-specific keywords.

All Programs Page

The All Programs page lists all known Air Force programs so that the user can quickly see which Air Force programs have a contractor or subcontractor with an elevated relative risk while providing a quick overview of the riskiest contractors working on each of the Air Force programs. Figure 4.5 shows a snapshot of the page with notional data. Programs are ordered by the degree of risk of each contractor who works on them. Acquisition personnel can drill down on each row in this table to reach a webpage describing the specific state of that program.

Figure 4.5. Prototype User Interface: All Programs Page

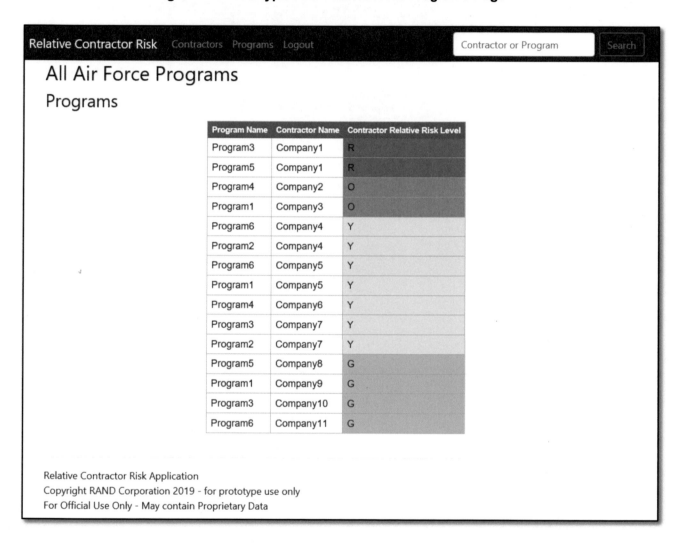

Single Program Page

The Single Program page lists a series of relevant details for a single specific Air Force program. Figure 4.6 shows a snapshot of the page with notional data. First, it lists any known prime contractors for this program, derived from FPDS and PMRT data. It also lists the calculated overall relative risk level for those contractors. Next, it displays all known subcontractors on this program, displaying the calculated overall relative risk level for each subcontractor as well. Finally, it lists the three PAR program evaluation scores for programs where PAR data are available. Each of the rows in the "Prime Companies" and "Subcontractors" tables allows the user to drill down into the details of that contractor's risk factors; clicking on a row in the table will take the user to the appropriate Single Contractor page.

Figure 4.6. Prototype User Interface: Single Program Page

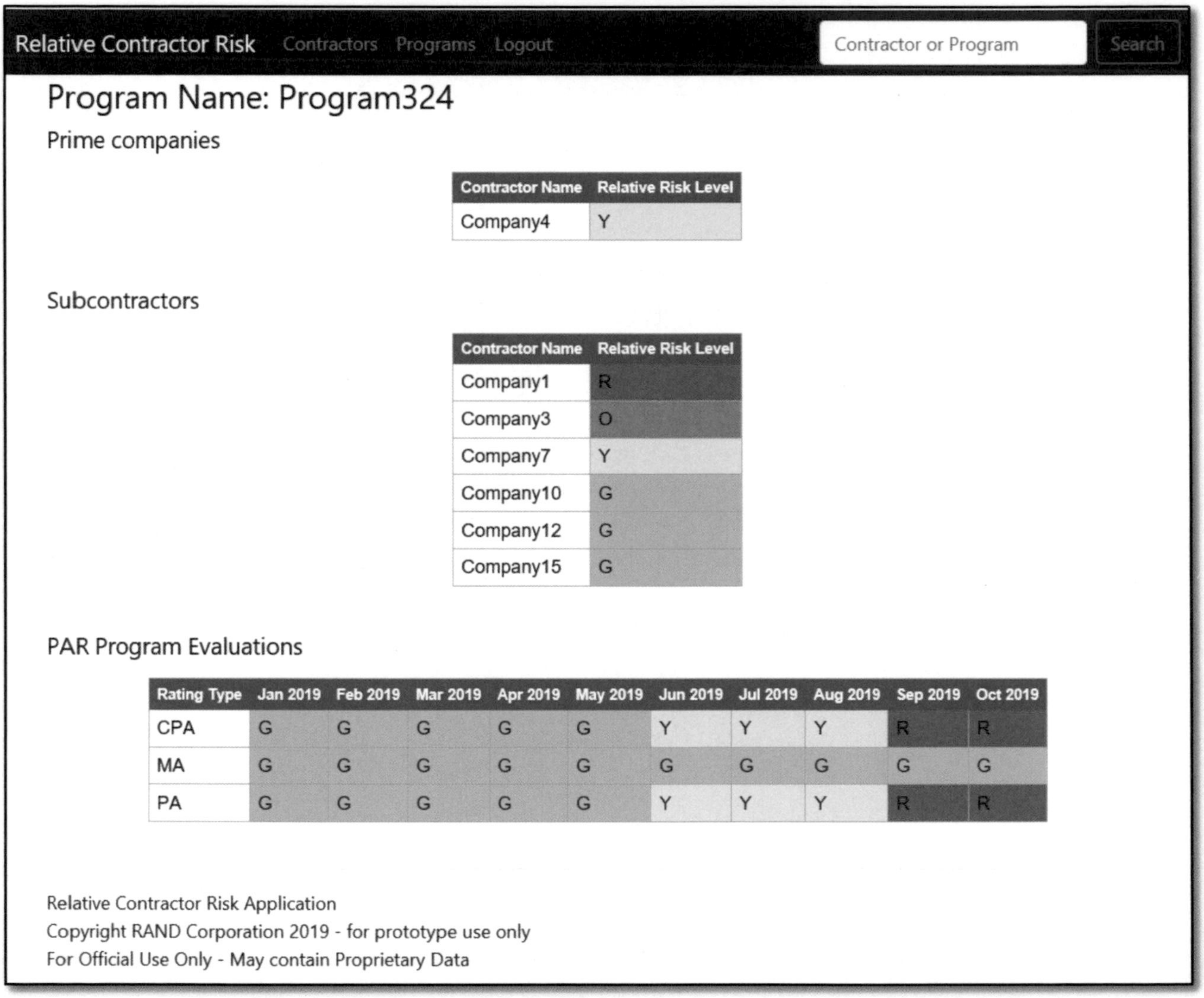

Relative Contractor Risk	Contractors Programs Logout		Contractor or Program	Search

Program Name: Program324

Prime companies

Contractor Name	Relative Risk Level
Company4	Y

Subcontractors

Contractor Name	Relative Risk Level
Company1	R
Company3	O
Company7	Y
Company10	G
Company12	G
Company15	G

PAR Program Evaluations

Rating Type	Jan 2019	Feb 2019	Mar 2019	Apr 2019	May 2019	Jun 2019	Jul 2019	Aug 2019	Sep 2019	Oct 2019
CPA	G	G	G	G	G	Y	Y	Y	R	R
MA	G	G	G	G	G	G	G	G	G	G
PA	G	G	G	G	G	Y	Y	Y	R	R

Relative Contractor Risk Application
Copyright RAND Corporation 2019 - for prototype use only
For Official Use Only - May contain Proprietary Data

50

5. Insights, Conclusions, and Next Steps

Insights on Data Utility and Availability

This project aimed to combine traditional and nontraditional[1] data to provide acquisition professionals earlier indications of relative contractor performance risks. Although there are several traditional ways that program information is available (e.g., MARs, SARs, DAES), supplementing these with other data sources—government, public, and commercial—is useful. There are some data that could potentially make the relative risk comparisons more robust but are not available to the public and require specific permissions from the requisite authorities. Although some of these data fall into the Sensitive Data category as authorized for release to FFRDCs under Section 235 of the FY 2017 National Defense Authorization Act,[2] there are additional useful data sources that remain unavailable.

Sensitive Data Access and Use

Some data that could be very useful for assessing relative contractor risk performance are very difficult to gain access to, even for FFRDCs and probably for government officials. CPARS is an example. It records recent contractor performance on all federal contracts, depending on the business sector and dollar thresholds as seen in Table 5.1. This covers Acquisition Category (ACAT) I–III programs.

---

[1] Again, these data are nontraditional for government PMs. These data are used by a great many people for other purposes.

[2] Public Law 114-328, Section 235 (G) (2016), defines *sensitive information* as

> confidential commercial, financial, or proprietary information, technical data, contract performance, contract performance evaluation, management, and administration data, or other privileged information owned by other contractors of the Department of Defense that is exempt from public disclosure under section 552(b)(4) of Title 5, United States Code, or which would otherwise be prohibited from disclosure under section 1832 or 1905 of Title 18, United States Code (Public Law 114-328, 2016).

This law authorized a pilot program permitting the release of such information to FFRDCs. This research project was a recipient of this kind of information.

Table 5.1. CPARS Reporting Thresholds

Business Sector	Dollar Threshold[a]	Reviewing Official[b]
DoD services and agencies		
Systems (includes new development and major modifications)	> $5,000,000	One level above the PM[c]
Nonsystems		
Operations support	> $5,000,000[d]	One level above the assessing official
Services	> $1,000,000	One level above the assessing official
Information technology	> $1,000,000	One level above the assessing official
Ship repair and overhaul	> $500,000	One level above the assessing official
Architect-engineer	≥ $35,000; all terminations for default	One level above the assessing official
Construction	≥ $700,000; all terminations for default	One level above the assessing official

SOURCE: CPARS Project Manager, 2020, p. 24.

[a] "The contract/order thresholds for performance collection [see FAR 42.1502 (2020)] apply to the 'aggregate' value of contracts/orders; that is, if a contract's/order's original award value were less than the applicable threshold but subsequently the contract/order was modified and the new value is greater than the threshold, then evaluations are required to be made, starting with the first anniversary that the contract's/order's face value exceeded the threshold. If the total contract/order value including unexercised options and orders (for [indefinite delivery/indefinite quantity] contracts, total estimated value of unexercised options and orders) is expected to exceed the collection threshold, initiate the collection process at the start of the contract/order. Buying activities may choose to collect performance evaluations for awards below these thresholds." (CPARS Project Manager, 2020, p. 24)

[b] "Only required when the contractor indicates nonconcurrence with the evaluation or if otherwise requested by the contractor during the 60-calendar day comment period." (CPARS Project Manager, 2020, p. 24)

[c] "(Or equivalent individual) responsible for program, project or task/job order execution." (CPARS Project Manager, 2020, p. 24)

[d] "For contracts/orders under the reporting thresholds, buying activities should continue to accumulate contractor performance data from existing management information systems, which already capture data on timeliness of delivery and quality of product or service." (CPARS Project Manager, 2020, p. 24)

Other valuable data are in DCMA PARs. These provide an internal government assessment of a contractor's performance on a program at one or more locations. It includes assessments of contractor performance, management, and production. This information is typically closely held and is marked proprietary.

The Air Force provided or sponsored access to other government-controlled data to include PMRT, AIR, DAVE, and more via the team's participation in the Section 235 pilot. These data were used to identify program contracts and link contractors to programs and were used to compare a contractor with its peers to assess the relative contractor performance risks in the categories in Table 1.1.

Financial Data

Company financial data are important in assessing the company's health, ability to absorb and successfully execute work, and development of technical prowess. These data are available from the federal government through the SEC'S EDGAR database and from commercial data providers (i.e., FactSet, Eikon, and S&P Global). Although it is possible to create financial histories from SEC filings, this is very labor-intensive because the annual filings, even though they have somewhat stable formats, can vary within a company between filings and also between companies. Also, these filings are typically unstructured data and would require scraping from PDFs and cleaning to be useful. The commercial providers provide data in a structured manner and provide the capability to determine relationships between companies (e.g., supply chain). Financial reports are provided as structured data that can be analyzed over years to determine trends, norms, deviations, and more, and also compare companies with their peers to calculate relative risk. Consequently, the research team selected a commercial provider, S&P Global.[3]

Jobs Postings

As part of the workforce relative risk calculation, the research team purchased data from Monster Worldwide, Inc., which owns and operates Monster.com, a global employment website.[4] These data were used to understand the types of skills that were required at a work location for a particular contract. Understanding how long a position was vacant provides an indication of how difficult it is to fill. Highlighting this to the government PM or oversight committee allows him or her to assess the criticality of this position to the program of interest and apply the necessary focus.

Practical Implementation Insights

Once we had access to data, there were some practical challenges we resolved for the algorithms to make sense and for the information to be displayed in a coherent manner. Two of the most challenging areas are associating company information due to differences in nomenclature and changes across time. Another challenge is how to display the information in an intuitive manner for acquisition professionals.

Correlating Companies: Name Differences

One issue the research team confronted is the lack of a consistently used, unique identifier for companies within the available data. As an example, Boeing could be listed in several ways: "The Boeing Company"; "Boeing Company, The"; "Boeing Company"; or simply "Boeing." Couple this with potential spelling errors, and it becomes more difficult to ensure that proper correlation of data is occurring. Therefore, it is necessary to develop an approach that allows character-matching to link company spellings accordingly.

---

[3] The research team also looked at Eikon and FactSet as potential providers of financial data. These two providers could have also been used instead of S&P Global; however, the Air Force was already familiar with S&P Global and had used these data in the past.

[4] The research team used Monster.com for its global reach and terms of data use. However, there are other job websites that could also have worked (e.g., Indeed.com and ihire.com).

An additional wrinkle are joint ventures. These combinations of companies typically do not have assets assigned to them but are legal entities where the components, i.e., company partners, retain the resources. When this occurs, it is difficult to properly attribute financial information or programs appropriately.

Some data sources, such as SAM.gov and FPDS-NG, use DUNS numbers. The nine-digit DUNS numbers are unique to a company and are assigned at the lowest organizational level, such as specific business locations with unique, separate, and distinct operations (Dun & Bradstreet, undated). These data sources also include contract numbers, which allows linkages to programs.

Correlating Companies: Across Time

Various actions can make it difficult to correlate companies across time, including M&As, reorganizations of business units, locational moves and consolidations, and others. As companies change names or combine with other companies, their financial histories are less easily traceable—therefore, metrics that depend on trend information are more difficult to calculate.

Visualization Approach

As acquisition professionals have limited time, showing these relative contractor risk indicators is critical for this tool to be useful. At this time, the team is using a color scheme familiar to the acquisition community—green, yellow, and red—and have added orange to indicate an intermediate degradation between yellow and red.[5] The web interface was designed to highlight the negatives and to provide the user with changes in status over time at a high level. This would focus the acquisition professional on areas where problems might manifest and to see whether relative risks are degrading.

Conclusions

Our research shows that it is possible to access and combine traditional and nontraditional data sources to highlight the areas of additional management attention for the government. It would then be up to the government acquisition professional to use his or her program insights to determine whether this focus area adversely affects the program or raises risks to a level no longer acceptable to the government. However, this is just the beginning. Additional data and the inclusion of more metrics are necessary to make this more robust; more measures are needed in each of the 11 risk categories for a balanced approach. For this to be successful, data availability and accessibility are key. Also, to reduce errors induced by manual manipulation of data, API or machine-to-machine access is preferred. This makes it easier to handle large data sets more efficiently.

---

[5] No intermediate color between green and yellow was added because anything not green merited attention.

The algorithms can be further "tuned" as more data become available. Longer periods of data and more-diverse sources of data about the same underlying risk condition could provide increasing amounts of context in which to examine a particular risk condition and allow more fine-grained filtering of potential contractor risks.

Additionally, it is important to understand the types of questions the user communities will be trying to answer. A PM is typically focused on one program with one contract team. Consequently, a PM might focus on the Program page almost exclusively. A program executive officer or MDA has a portfolio to manage and might find the All Programs page more useful, with its ability to drill down into specifics about a program's contractor team as needed. Other officials with oversight responsibilities, i.e., performance of the defense industrial base, might find the All Contractors page view more interesting because it could provide insights into issues that different segments of the industrial base are having.

The risk measures discussed in this report combine available data to indicate when a company may have an increased risk when compared with the values for other companies. Although the approach usually cannot definitively identify that a risk exists, those measures are objectively applied to all companies and constitute a way to reduce the number of issues for further due diligence. We have an initial working prototype, but further work is needed to evaluate the sensitivity and utility of the proposed measures once the prototype is matured and more data are added.

Next Steps

As indicated earlier, this is an initial report on work in progress. In addition to the improvement options discussed in Chapter 3 for each relative risk measure algorithm, we observed the following potentials for improvement in identifying relative contractor risks.

Gaining Access to Other Data

There are other data sets that could be immediately useful. Access to all government contracts through EDA would allow for comparisons across contracting types, incentive schemes, implementation of clauses, and more to understand their impacts. This would have acquisition and contracting policy implications.

Access to CPARS and integrating these data into the tool would make the assessment of contractor performance more robust. It would also make this valuable data source more available to the acquisition community and could expand its use.

DCMA is the organization charged with monitoring contractor execution performance and, therefore, has complied data for many programs of interest. The PARs are one such instance of the kind of available data. Incorporating other DCMA data into this approach would also enhance the insight of acquisition professionals and help with earlier notification of contractor performance risks.

Within the current cybersecurity environment, it would be helpful for acquisition professionals to have an increased understanding of their contractor teams' security posture. This directly affects program security and safety. Therefore, properly making this more available would benefit programs.

Implementing Other Possible Risk Measures Identified

As discussed in Chapter 2, our risk taxonomy identified many more potential risk measures than we were able to implement in the prototype because of resource and data access limitations. The following is one example risk measure in the cost category that we examined but did not yet implement.

Inflation in Principal Place of Performance

Motivation. Here we would check whether inflation is relatively high in the principal place of performance (for an example of regional inflation differences, see the Federal Reserve Bank of St. Louis, 2017). If so, then the government might expect contract costs (prices paid) to increase above normal. This would be mitigated on firm fixed-price contracts, but subsequent contract costs (e.g., for a subsequent production lot) could be higher.

Approach. For each DoD contract, the system would use the reported principal place of performance in FPDS-NG. The system would then look up the BLS regional inflation in which the principal place of performance is located. Relative cost risk would be rated based on some statistical measure of the difference in inflation from the national average.

Improvement options. Ideally, we would compare job-specific unemployment rates in the area with the primary PSC for the contract work (or better yet, contractor shortfalls by job). These improvements may be considerations for future upgrades (if appropriate data can be accessed).

Expanding from Early Identification of Risks to Early Identification of Execution Problems

A logical next step that complements additional relative contractor performance risk indicators is coupling that with program execution risks. PMs seek early indicators of program performance problems to enable proactive mitigation. Contractors and government overseers produce various reports and data, but the resulting volume of information is too large for PMs and supervisors to manually monitor and pick out early indicators of actual execution problems. Program support staff are tasked to monitor these data, but these efforts can be time-consuming and distract from higher-level tasks to proactively address and correct these issues.

Text analysis tools can improve search capabilities to locate known topical indicators of problems in free text through preprocessing of information and facilitating search by keyword, context, and (in some cases) sentiment and tone. Applying these to Air Force program management requires both access to internal government data and acquisition expertise.

Although acquisition experts know of some indicators of program performance, early indicators are few. Performance indicators in core data, such as the SARs, tend to be trailing or limited. Artificial intelligence (e.g., machine learning or neural networks) and traditional statistical analysis tools applied to new data sources (e.g., contractor and government reports in DCMA databases) may be able to identify new leading indicators.

Some of this information is already available via existing data sources, e.g., PMRT, where acquisition professionals currently report program execution information. Combining these views could provide acquisition professionals with insights into how some contractor risks manifest in program execution.

Appendix A. Risk Measures and Potential Data Sources

Our full taxonomy included many more relative risk measures than we were able to obtain data for and implement in the initial prototype. To help the reader understand the breadth of the measures we considered, we developed this appendix. Its sole intent is to provide the reader with the measures we considered, required data, measurement approach, and algorithm development difficult. Using the information parameters defined in Table A.1, we briefly describe these measures and potential data sources in Table A.2 through Table A.13. Each risk measure has one or more inputs and associated data; each are shown on separate rows.

Table A.1. Information Provided on Risk Measures

Information Type	Description
Name	A descriptive title for each specific measure; similar risk measures (e.g., different measures of hiring risk) are differentiated with parenthetical or augmented information that indicates the key data driver used in the measure
Input(s)	Conceptual descriptions of the inputs to the measure calculation
Data and possible source	Specific data proving the inputs and possible sources for these data
Approach notes	Brief comments about the approach for combining the data to inform the relative risk
Algorithm difficulty	An approximate qualitative assessment of the difficulty in implementing the algorithm related to this input portion of the risk measure

Name	Input(s)	Data and Possible Source	Approach Notes	Algorithm Difficulty
Hiring risk (hiring needed): general	Hiring need	Staff shortfall	Use staffing plan from proposal; use company postings; use media stories about job openings.	High
	Unemployment rate in location	Unemployment rate (primary work location, notes)	Use BLS to get CPI for unemployment rate in location.	Low
Hiring risk (hiring needed): cleared staff	Hiring need for cleared personnel	Staff shortfall	MAR would have in comments or risks; search staffing plan in proposal; look at postings for location of work.	High
Hiring risk (hiring needed): slow start (EV)	Hiring needs: execution trends	Level 1 EV data	Get EV data (CADE first choice), compare shape with theoretical Rayleigh Function prediction (e.g., see Davis, Christle, and Abba, 2009).	High
Hiring risk (hiring needed): slow start (expenditures)	Hiring needs: contract expenditure trends	Expenditures by contracts	Compare with Rayleigh Function for contract expenditures; display risk at program level keeping underlying data.	High
Hiring risk (unemployment): primary work location	Primary work location	Primary work location	Start with FPDS; if EDA is available, then use as primary.	Low
	Unemployment rate in primary work location	Unemployment rate (primary work location, notes)	Use BLS to get CPI for unemployment rate in location.	Low
Hiring risk (attractiveness): primary work location	Primary work location	Primary work location	Start with FDPS; if EDA is available, then use as primary.	Low
	Hiring attractiveness	BLS CPI; various rankings on the best places to work	Get work location from EDA/FPDS-NG; CPI from BLS; option is "Best Places to Work" for intangibles.	Low
Experience in key work area: Prior Fed Work	Contract number	PMRT		Low
	Contract topic area	PSC; tech domain	PSC code in FPDS; other data sources are optional.	Low
	Prior federal work	PSC	FPDS; CPARS for past performance rating.	Low

Name	Input(s)	Data and Possible Source	Approach Notes	Algorithm Difficulty
Experience in key work area: leadership experience	Contract number	PMRT		Low
	Contract topic area	PSC; tech domain	PSC code in FPDS; other data sources are optional.	Low
	Topic area to field	TBD		Medium
	Workforce experience	Leadership experience	Key personnel in proposal; corporate website for bios.	Medium
Experience in key work area: staff experience	Contract number	PMRT		Low
	Contract topic area	PSC; tech domain	PSC code in FPDS; other data sources are optional.	Low
	Topic area to field	TBD		Medium
	Workforce experience	Staff experience; degrees	Key personnel in proposal; corporate website for bios; jobs for fill requirements.	High
Experience in key work area: hiring need	Hiring need	Staff shortfall	Use staffing plan from proposal; use company postings; use media stories about job openings.	High
R&D spending trend (design team risk)	Contract number	PMRT		Low
	Contract topic area	PSC; tech domain	PSC code in FPDS; other data sources are optional.	Low
	Topic area to field	TBD		Medium
	Federal research, development, test, and evaluation (RDT&E) contract work trends	Obligations by R&D PSC	FPDS for PSC; sort by appropriation.	Low
	Independent R&D (IR&D) trends	DTIC IR&D database (also commercial sources)		Medium
Retention risk: salary growth	Contract topic area	PSC; tech domain	PSC code in FPDS; other data sources are optional.	Low
	Topic area in high demand	Salary growth	Identify topic area, search for changes in salary; look for retention trends.	Low
Hiring and retention risks: job openings	Contract topic area	PSC; tech domain	PSC code in FPDS; other data sources are optional.	Low
	Topic area in high demand	Job openings	Identify topic area, search for job openings; look for retention trends.	Low

NOTE: TBD = to be determined (as in more work is needed to identify).

Table A.4. Risk Measures: Cost

Name	Input(s)	Data and Possible Source	Approach Notes	Algorithm Difficulty
Inflation in primary work location(s)	Inflation in area	BLS CPI	Get local inflation rate (notify PM).	Low
	Primary work location	Primary work location	Start with FDPS; if EDA is available, then use as primary.	Low
	Contract number	PMRT		Low
Invest to increase capacity: jump in revenue	Large changes in nominal revenue trends	Revenue	Revenue from CapIQ.	Low
Declining revenue (overhead risk)	Revenue trends holding overhead constant	Revenue	Revenue from CapIQ.	Low
	Contract type	Contract type	FPDS is primary source.	Low
Production cost risk	Past cost growth	CSDRs (prior lots)	See the actual cost of prior lots and compare with current lot.	Medium
	Production line maturity	Stability of learning curve; placement on curve	Delivered quantities in SARs; calculate learning curve and look at R^2 (only ACAT I).	Medium
	Production quantity stability	Quantity trends	PMRT provides quantity baseline; compare changes over time.	Medium
Material/supply inflation	Key materials (raw materials and parts made from them)	Cost Analysis Requirements Description (CARD); independent cost estimates or Service cost positions	Cost documentation from AIR (CARD) or CADE (CARD); materials from contractor proposal.	Medium
	Inflators (original and current)	CARD; independent cost estimates or service cost positions; OSD's Green Book; others	Cost documentation from AIR (CARD) or CADE (CARD).	Medium
	Inflators (behavior)	CARD; independent cost estimates or service cost positions; OSD's Green Book; others	Cost documentation from AIR (CARD) or CADE (CARD).	Medium
	Contract type	Contract type	Identify contract type; focus on nonfixed price contracts. These are more susceptible to government cost risk. (FAR 16.2, Fixed Price Contracts; FAR 16.3, Cost-Reimbursable; FAR-16.4, Incentive Contracts).	Low

Table A.5. Risk Measures: Financial

Name	Input(s)	Data and Possible Source	Approach Notes	Algorithm Difficulty
Financial metrics: overall	Overall relative credit health score	CapIQ Credit Health Panel (combines operational, solvency, and liquidity scores)	Financial vendors pull from Moody's et al. at company (versus parent) level.	Low
Declining profits	Profit trends	CapIQ profit	Revenue from CapIQ.	Low
Financial metrics: operational	Operational relative health	CapIQ Credit Health Panel	Financial vendors pull from Moody's et al. at company level.	Low
Financial metrics: solvency	Solvency relative health	CapIQ Credit Health Panel	Extract at company (versus parent) level.	Low
Financial metrics: liquidity	Liquidity relative health	CapIQ Credit Health Panel	Extract at company (versus parent) level.	Low
Customer base (option 1)	Revenue trends	Company overall revenue; obligations and expenditures by contract	Revenue from a financial data vendor; contract obligations and expenditures from PMRT.	Low
Customer base (option 2)	Primary customers	Primary customers; federal customers	Get total federal spend from FPDS and financial data vendor; use PMRT information to compare program spend size.	Low

Table A.6. Risk Measures: Corporate Stability

Name	Input(s)	Data and Possible Source	Approach Notes	Algorithm Difficulty
Recent/pending M&A	Recent/pending M&A	M&A	Financial vendor data stream; hear from media; must be in annual reports.	Low
Excluded party (not allowed to bid on federal contracts)	Excluded companies not allowed to bid on federal contracts	SAM.gov excluded parties list	Look up the company in SAM.gov and see whether it is "excluded."	Low
Declining stock price	Stock price trends	CapIQ stock price	Revenue from CapIQ.	Low
Management turnover (C-suite)	Management stability	List of key management/issues	C-suite changes reported in media, press releases, and corporate website.	Low
Pending lawsuits	Active	Lawsuits (against the corporation or its key leaders)	Media reporting, court filings, annual reports.	Low
Lawsuit losses	Lawsuits lost	Lawsuits	Media reporting, court filings, annual reports.	Low

Table A.7. Risk Measures: Supply Chain

Name	Input(s)	Data and Possible Source	Approach Notes	Algorithm Difficulty
Supplier risk: Corporatewide suppliers	<risks: recursive>			
	Suppliers	CapIQ Suppliers List	Primary data source is financial data vendors.	Low
Supplier risk: contract-specific suppliers	<risks: recursive>			
	Suppliers	FSRS via USAspending.gov	Primary data source is financial data vendors.	Low
	Contract number	PMRT		Low
Insufficient/new supply chain (to include limited sources because of consolidations)	Contract number	PMRT		Low
	Supply chain for specific contract	Proposal	Primary data source to identify suppliers is financial data vendor; use proposal to help identify any "one-of-a-kind" discriminators that are critical to program execution.	Medium
Beyond prime suppliers	Dominance of primes in major components	CSDRs; FSRS	FSRS for contract awards; FPDS by PSC.	
Supplier quality assurance	Contract number	PMRT		Low
	Suppliers	FSRS or USAspending.gov	Primary data source is financial data vendors.	Low
	Supplier quality assurance	Supplier quality assurance	Must have DCMA data to attempt eTool record; might have information in the PAR; might have a CPARS reference if it were once a prime supplier and had a significant quality issue reported to the contracting officer.	High
Supplier part tampering	Supplier part tampering	Supplier part tampering	Recommend combining part tampering and counterfeit parts.	TBD
	Suppliers	Supply chain	Recommend combining part tampering and counterfeit parts.	TBD
Supplier counterfeit parts	Supplier counterfeit parts	Supplier counterfeit parts	Recommend combining part tampering and counterfeit parts.	TBD
	Suppliers	Supply chain	Recommend combining part tampering and counterfeit parts.	TBD

NOTE: TBD = to be determined.

Table A.8. Risk Measures: Government Influence

Name	Input(s)	Data and Possible Source	Approach Notes	Algorithm Difficulty
Air Force is a significant customer	Primary customers	Primary customers; federal customers	FPDS and financial data vendors for revenue; use PMRT for contract obligations and expenditures.	Low
	Revenue	Company revenue; Air Force obligations and expenditures	CapIQ for total revenue. Compare with total obligations in PMRT or FPDS-NG.	Low
DoD is a significant customer	Primary customers	Primary customers; federal customers	CapIQ for total revenue. Compare with total obligations in PMRT or FPDS-NG.	Low
	Revenue	Revenue; obligations; expenditures;	CapIQ for total revenue. Compare with total obligations in PMRT or FPDS-NG.	Low
Federal government is a significant customer	Primary customers	Primary customers; federal customers	FPDS and financial data vendors for revenue; use PMRT for contract obligations and expenditures.	Low
	Revenue	Revenue; obligations; expenditures;	FPDS and financial data vendors for revenue; use PMRT for contract obligations and expenditures.	Low

64

Table A.9. Risk Measures: Contractor Performance

Name	Input(s)	Data and Possible Source	Approach Notes	Algorithm Difficulty
Prior work experience in this product/service	Contract topic area	PSC; tech domain	Search FPDS by PSC, examine proposal for past experience, look for DCMA assessments.	Low
	Prior contract topic areas	PSC; tech domain	Search FPDS by PSC, examine proposal for past experience, look for DCMA assessments.	Low
Prior DoD experience	Prior contracts	DoD contracts	Search FPDS by PSC, examine proposal for past experience, look for DCMA assessments.	Low
	(a) Provenance of company (what consolidations have occurred)	DUNS history from SAF/AQX		Low
	or (b) M&A history	M&A	Fin vendor, SEC reporting (legal requirement), media reporting for large mergers.	Low
Prior federal experience	Prior contracts	Federal contracts	Search FPDS by PSC, examine proposal for past experience, look for DCMA assessments.	Low
	(a) Provenance of company (what consolidations have occurred)	DUNS history from SAF/AQX		Low
	or (b) M&A history	M&A	Fin vendor, SEC reporting (legal requirement), media reporting for large mergers.	Low
Past/current government performance: PARs	Ability to execute	Contract past performance in DCMA PARs	Lexical analyses of documents; some numerical structure to PAR or corrective action reports.	Medium
Past/current government performance: CPARS	Ability to execute	Contract past performance in CPARS	Lexical analyses of documents; some numerical structure to PAR or corrective action reports.	Medium
Commercial reputation	Management attention	Media reporting	Examine news articles/feeds; look at fin vendor summaries/articles.	Medium

Table A.10. Risk Measures: Security

Name	Input(s)	Data and Possible Source	Approach Notes	Algorithm Difficulty
Recent cyber compromises	Cyber compromises	Cyber compromises	Text search for media reporting and press releases.	High
	System security	Program protection plan	Typically, a government document; search in AIR as primary.	Low
Meeting new DoD cyber requirements	Cybersecurity	Unknown	Systems are being built, but access, appropriateness, and usability are unknown.	High
Sufficient cleared staff	Number and levels of cleared staff needed	Unknown	Best source would be the proposal; may not be required to state a definitive number.	TBD
	Number and levels of cleared staff needed	Unknown	Best source would be the proposal; may not be required to state a definitive number.	TBD
Sufficient cleared workspace	Cleared workspace	Unknown	DSS; company cage code; addressed in proposal.	TBD
	Cleared workspace needed	Unknown	RFP provides need; proposal addresses; DSS confirms.	TBD

NOTE: TBD = to be determined.

Table A.11. Risk Measures: Infrastructure

Name	Input(s)	Data and Possible Source	Approach Notes	Algorithm Difficulty
Production stability	Contract number	PMRT		Low
	Production quantity stability	Quantity trends	Quantity changes in MAR; issues in DCMA reporting.	Low
New infrastructure	Contract number	PMRT		Low
	New infrastructure needed	Proposals (discussion; non-recurring engineering)		Medium

Table A.12. Risk Measures: Capacity

Name	Input(s)	Data and Possible Source	Approach Notes	Algorithm Difficulty
Production capacity: EOQ	Contract number	PMRT		Low
	Production capacity	Production capacity: EOQ	Look for EOQ (minimum and maximum) reporting in MAR (or SAR) and look at PAR for any production issues.	Low
Jump in corporate revenue (capacity risk) *[similar to item in cost category]*	Revenue trends	Revenue	Revenue from CapIQ.	Low

Table A.13. Risk Measures: Future

Name	Input(s)	Data and Possible Source	Approach Notes	Algorithm Difficulty
Low R&D investments	R&D trends	CapIQ (SEC filings)		Low
Low government-funded R&D	Contract topic area	PSC; tech domain	Search FPDS by PSC, examine proposal for past experience, look for DCMA assessments.	Low
	Recent RDT&E trends in contract topic areas	FPDS-NG	Look up RDT&E funding in the PSC for the company.	Low
Low publication rates in area of interest (Bibliometrics: DTIC Reports)	Contract topic area	PSC; tech domain	Search FPDS by PSC, examine proposal for past experience, look for DCMA assessments.	Low
	Recent RDT&E trends in contract topic areas	DTIC tech reports and DoDTechipedia	Look up RDT&E funding in the PSC for the company.	Medium
Low publication rates in area of interest (Bibliometrics: Web of Science)	Contract topic area	PSC; tech domain	Search FPDS by PSC, examine proposal for past experience, look for DCMA assessments.	Low
	Recent RDT&E trends in contract topic areas	Web of Science	Look up RDT&E funding in the PSC for the company.	Medium
Low patent rates in area of interest (Patentometrics)	Contract topic area	PSC; tech domain	Search FPDS by PSC, examine proposal for past experience, look for DCMA assessments.	Low
	Recent RDT&E trends in contract topic areas	Web of Science	Look up RDT&E funding in the PSC for the company.	High
Low IR&D investments	IR&D trends	DTIC IR&D database (potentially commercial sources)		Medium
Declining revenue			*(see calculation under the Cost category in Table A.4)*	

68

References

Alteryx, homepage, undated. As of December 27, 2019:
https://www.alteryx.com

Barnett, Jackson, "DOD Unveils Plans for Contractor Cybersecurity Standards," *FedScoop*, June 14, 2019. As of November 5, 2020:
https://www.fedscoop.com/dod-cmmc-new-supply-chain-and-cybersecurity-rules/

BLS—*See* U.S. Bureau of Labor Statistics.

Bradshaw, James, and Su Chang, "Past Performance as an Indicator of Future Performance: Selecting an Industry Partner to Maximize the Probability of Program Success," *Defense Acquisition Research Journal*, Vol. 20, No. 1, April 2013, pp. 59–80. As of December 27, 2019:
https://www.dau.edu/library/arj/_layouts/15/WopiFrame.aspx?sourcedoc=/library/arj/ARJ/A RJ65/ARJ_65-Bradshaw.pdf&action=default

Bolten, Joseph G., Robert S. Leonard, Mark V. Arena, Obaid Younossi, and Jerry M. Sollinger, *Sources of Weapon System Cost Growth: Analysis of 35 Major Defense Acquisition Programs*, Santa Monica, Calif.: RAND Corporation, MG-670-AF, 2008. As of January 2, 2020:
https://www.rand.org/pubs/monographs/MG670.html

Bounds, Thomas, Andrew Clark, Todd Henry, John Nierwinski, Suzanne Singleton, and Brian Wilder, *Army Independent Risk Assessment Guidebook*, Aberdeen Proving Ground, Md.: U.S. Army Materiel Systems Analysis Activity, TR-2014-19, April 2014. As of July 31, 2019:
https://apps.dtic.mil/dtic/tr/fulltext/u2/a599874.pdf

Chernick, Michael R., *The Essentials of Biostatistics for Physicians, Nurses, and Clinicians*, Hoboken, N.J.: John Wiley & Sons, 2011.

Cousins, Paul D., Richard C. Lamming, and Frances Bowen, "The Role of Risk in Environment-Related Initiatives," *International Journal of Operations & Production Management*, Vol. 24 No. 6, 2004, pp. 554–565. As of December 26, 2019:
https://doi.org/10.1108/01443570410538104

CPARS Project Manager, *Guidance for the Contractor Performance Assessment Reporting System (CPARS)*, Portsmouth, N.H.: U.S. Department of Defense, October 2020. As of December 10, 2020:
https://www.cpars.gov/documents/CPARS-Guidance.pdf

Davis, Dan, Gary Christle, and Wayne Abba, *Using the Rayleigh Model to Assess Future Acquisition Contract Performance and Overall Contract Risk*, Arlington, Va.: Center for Naval Analysis, CRM D0019289.A2/Final, January 2009.

Deeb, George, "7 Potential Pitfalls With Mergers & Acquisitions," *Forbes*, November 2016. As of December 27, 2019:
https://www.forbes.com/sites/georgedeeb/2016/11/02/7-potential-pitfalls-with-mergers-acquisitions/#662b06e6170f

Deloitte, *Cultural Issues in Mergers and Acquisitions*, New York: Deloitte Development LLC, 2009. As of December 27, 2019:
https://www2.deloitte.com/content/dam/Deloitte/us/Documents/mergers-acqisitions/us-ma-consulting-cultural-issues-in-ma-010710.pdf

Department of Defense Instruction 8510.01, *Risk Management Framework (RMF) for DoD Information Technology (IT)*, Incorporating Change 2, Washington, D.C.: U.S. Department of Defense, March 12, 2014, effective July 28, 2017. As of July 31, 2019:
https://www.esd.whs.mil/Portals/54/Documents/DD/issuances/dodi/851001p.pdf?ver=2019-02-26-101520-300

Deputy Assistant Secretary of Defense for Systems Engineering, *Department of Defense Risk, Issue, and Opportunity Management Guide for Defense Acquisition Programs*, Washington, D.C.: U.S. Department of Defense, 2017.

DoD—*See* U.S. Department of Defense.

Domo, homepage, undated. As of December 27, 2019:
https://www.domo.com/

Dun & Bradstreet, "About the D-U-N-S Number," webpage, undated. As of July 31, 2019:
https://fedgov.dnb.com/webform/pages/dunsnumber.jsp

Dundas BI, homepage, undated. As of December 27, 2019:
https://www.dundas.com/

FAR—*See* Federal Acquisition Regulation.

Federal Acquisition Regulation 42.1502 Policy, effective as of October 26, 2020. As of November 10, 2020:
https://www.acquisition.gov/far/42.1502

Federal Contractor Registry, "SAM & The Excluded Parties List: What Does It Mean?" webpage, January 30, 2019. As of July 25, 2019:
http://federalcontractorregistry.com/sam-the-excluded-parties-list-what-does-it-mean/

Federal Reserve Bank of St. Louis, "Regional Inflation," *The FRED® Blog*, webpage, January 5, 2017. As of July 20, 2019:
https://fredblog.stlouisfed.org/2017/01/regional-inflation/

Fitch Ratings, "Ratings Definition," webpage, undated. As of December 20, 2019:
https://www.fitchratings.com/site/definitions

iDashboards, homepage, undated. As of December 27, 2019:
https://www.idashboards.com/

InsightSquared, homepage, undated. As of December 27, 2019:
https://www.insightsquared.com/

Khan, Omera, and Bernard Burnes, "Risk and Supply Chain Management: Creating a Research Agenda," *International Journal of Logistics Management*, Vol. 18, No. 2, August 21, 2007, pp. 197–216. As of December 26, 2019:
https://doi.org/10.1108/09574090710816931

Lermack, Harvey B., *Steps to a Basic Company Financial Analysis*, Philadelphia, Pa.: Philadelphia University, May 2003.

Lorell, Mark A., Robert S. Leonard, and Abby Doll, *Extreme Cost Growth Themes from Six U.S. Air Force Major Defense Acquisition Programs*, Santa Monica, Calif.: RAND Corporation, RR-1761-AF, 2015. As of January 2, 2020:
https://www.rand.org/pubs/research_reports/RR630.html

Lorell, Mark A., Leslie Adrienne Payne, and Karishma R. Mehta, *Program Characteristics That Contribute to Cost Growth: A Comparison of Air Force Major Defense Acquisition Programs*, Santa Monica, Calif.: RAND Corporation, RR-1761-AF, 2017. As of January 2, 2020:
https://www.rand.org/pubs/research_reports/RR1761.html

Marine Corps Systems Command, *MARCORSYSCOM Acquisition Guidebook (MAG)*, Marine Corps Base Quantico, Va., last updated February 3, 2017.

Mayer, Lauren A., Mark V. Arena, and Michael McMahon, *A Risk Assessment Methodology and Excel Tool for Acquisition Programs*, Santa Monica, Calif.: RAND Corporation, RR-262-OSD, 2013. As of November 5, 2020:
https://www.rand.org/pubs/research_reports/RR262.html

McKernan, Megan, Nancy Young Moore, Kathryn Connor, Mary E. Chenoweth, Jeffrey A. Drezner, James Dryden, Clifford A. Grammich, Judith D. Mele, Walter T. Nelson, Rebeca Orrie, Douglas Shontz, and Anita Szafran, *Issues with Access to Acquisition Data and Information in the Department of Defense: Doing Data Right in Weapon System Acquisition*,

Santa Monica, Calif.: RAND Corporation, RR-1534-OSD, 2017. As of December 26, 2019:
https://www.rand.org/pubs/research_reports/RR1534.html

Merger Integration, "The 10 Most Common Post Merger Integration Problems," webpage, undated. As of November 5, 2020:
https://www.mergerintegration.com/10-common-post-merger-integration-problems

Mitchell, Vincent-Wayne, "Organizational Risk Perception and Reduction: A Literature Review," *British Journal of Management*, Vol. 6, No. 2, June 1995, pp. 115–133. As of January 21, 2019:
10.1111/j.1467-8551.1995.tb00089.x

Moody's Investor Service, *Rating Methodology Global Aerospace and Defense Industry*, New York, April 25, 2014.

Naval Air Systems Command, *NAVAIR Acquisition Guide 2016/2017*, Naval Air Station Patuxent River, Md., September 22, 2015.

Office of Acquisition, Analytics and Policy, "Assessments and Root Cause," webpage, Washington, D.C., U.S. Department of Defense, undated. As of January 2, 2020:
https://www.acq.osd.mil/aap/#/arc

Office of Management and Budget, *North American Industry Classification System*, Washington, D.C.: Executive Office of the President of the United States, 2017. As of July 30, 2019:
https://www.census.gov/eos/www/naics/2017NAICS/2017_NAICS_Manual.pdf

Office of the Under Secretary of Defense for Acquisition and Sustainment, *Department of Defense Earned Value Management Implementation Guide (EVMIG)*, Washington, D.C.: U.S. Department of Defense, January 18, 2019. As of January 2, 2020:
https://www.acq.osd.mil/evm/assets/docs/DOD%20EVMIG-01-18-2019.pdf

———, "Cybersecurity Maturity Model Certification," webpage, Washington, D.C.: U.S. Department of Defense, undated. As of July 29, 2019:
https://www.acq.osd.mil/cmmc/index.html

Office of the Under Secretary of Defense (Comptroller), *National Defense Budget Estimates for FY 2019*, Washington, D.C.: U.S. Department of Defense, April 2018. As of July 30, 2019:
https://comptroller.defense.gov/Portals/45/Documents/defbudget/fy2019/FY19_Green_Book.pdf

Oracle Solutions Business Intelligence, homepage, undated. As of December 27, 2019:
https://www.oracle.com/solutions/business-analytics/business-intelligence/

Parker, William, *Defense Acquisition University Program Managers Tool Kit*, 16th ed., Fort Belvoir, Va.: Defense Acquisition University, January 2011.

Porter, Gene, Brian Gladstone, C. Vance Gordon, Nicholas Karvonides, R. Royce Kneece, Jr., Jay Mandelbaum, and William D. O'Neil, *The Major Causes of Cost Growth in Defense Acquisition*, Volume I: *Executive Summary*, Alexandria, Va.: Institute for Defense Analyses, P-4531, 2009. As of January 2, 2020:
https://apps.dtic.mil/dtic/tr/fulltext/u2/a519883.pdf

Public Law 114-328, National Defense Authorization Act for Fiscal Year 2017, Section 235, Pilot Program on Disclosure of Certain Sensitive Information to Federally Funded Research and Development Centers, December 23, 2016. As of November 5, 2020:
https://www.congress.gov/114/plaws/publ328/PLAW-114publ328.pdf

RapidRatings, "Supplier Financial Risk: Health Assessment Report," webpage, undated. As of January 2, 2020:
https://www.rapidratings.com/resources/whitepapers/supplier-financial-risk-assessment-whitepaper/

Ritchie, Bill, and Clare Brindley, "An Emergent Framework for Supply Chain Risk Management and Performance Measurement," *Journal of the Operational Research Society*, Vol. 58, No. 11, 2007, pp. 1398–1411. As of December 27, 2019:
http://www.jstor.org/stable/4622835

S&P Global, "The S&P Capital IQ® Platform," brochure, New York, 2017. As of October 20, 2020:
https://www.spglobal.com/marketintelligence/en/documents/sp-global-capital-iq-platform-brochure.pdf

SAP, "SAP Crystal Reports: At a Glance," webpage, undated. As of December 27, 2019:
https://www.sap.com/products/crystal-reports.html?infl=05f03c7f-c3f8-4f62-8266-8f79fec692f2

SEC—*See* U.S. Securities and Exchange Commission.

Seth, Shobhit, "Top Reasons Why M&A Deals Fail," *Investopedia*, May 21, 2019. As of December 27, 2019:
https://www.investopedia.com/articles/investing/111014/top-reasons-why-ma-deals-fail.asp

Sisense, homepage, undated. As of February 19, 2021:
https://www.sisense.com/

Standard & Poor's, *Guide to Credit Rating Essentials*, New York: McGraw Hill Financial, July 2014. As of December 27, 2019:
https://www.spratings.com/documents/20184/760102/SPRS_Understanding-Ratings_GRE.pdf

Under Secretary of Defense for Acquisition, Technology, and Logistics, *Performance of the Defense Acquisition System: 2015 Annual Report*, Washington, D.C.: U.S. Department of Defense, 2015. As of January 2, 2020:
http://www.dtic.mil/docs/citations/ADA621941

———, *Performance of the Defense Acquisition System: 2016 Annual Report*, Washington, D.C.: U.S. Department of Defense, October 2016.

U.S. Bureau of Labor Statistics, "Quarterly Census of Employment and Wages," webpage, last updated June 12, 2019. As of November 5, 2020:
https://www.bls.gov/cew/classifications/area-guide.htm

U.S. Department of Defense, "The DoD Contractor Risk Assessment Guide," Washington, D.C., October 1988.

———, "Subject: Class Deviation—Micro-Purchase Threshold, Simplified Acquisition Threshold, and Special Emergency Procurement Authority," memorandum from the Under Secretary of Defense for Acquisition and Sustainment, Washington, D.C., August 31, 2018. As of August 1, 2019:
https://www.acq.osd.mil/dpap/policy/policyvault/USA002260-18-DPC.pdf

———, "Strategically Implementing Cybersecurity Contract Clauses," memorandum from Under Secretary of Defense for Acquisition and Sustainment, Washington, D.C., February 5, 2019. As of July 20, 2019:
https://www.acq.osd.mil/dpap/pdi/cyber/docs/USA000261-19%20USD%20Signed%20TAB%20A.pdf

U.S. General Services Administration, "GSA Forms Library: Form: SF1407, Pre-Award Survey of Prospective Contractor - Financial Capability," webpage, Washington, D.C., January 2014. As of October 14, 2020:
https://www.gsa.gov/forms-library/pre-award-survey-prospective-contractor-financial-capability

U.S. General Services Administration, Federal Acquisition Services, *Federal Procurement Data System Product and Service Codes Manual, August 2015 Edition*, Washington, D.C., October 1, 2015. As of July 20, 2019:
https://www.fpds.gov/downloads/top_requests/PSC_Manual_FY2016_Oct1_2015.pdf

U.S. Securities and Exchange Commission, "Updated Investor Bulletin: The ABCs of Credit Ratings," webpage, October 12, 2017. As of October 14, 2020:
https://www.sec.gov/oiea/investor-alerts-and-bulletins/ib_creditratings